Springer Natural Hazards

The Springer Natural Hazards series seeks to publish a broad portfolio of scientific books, aiming at researchers, students, and everyone interested in Natural Hazard research. The series includes peer-reviewed monographs, edited volumes, textbooks, and conference proceedings. It covers all categories of hazards such as atmospheric/climatological/oceanographic hazards, storms, tsunamis, floods, avalanches, landslides, erosion, earthquakes, volcanoes, and welcomes book proposals on topics like risk assessment, risk management, and mitigation of hazards, and related subjects.

Jürgen Herget

Flood Reconstruction

Palaeohydrological Approaches, Methods and Results

Springer

Jürgen Herget
Department of Geography
Bonn University
Bonn, Germany

ISSN 2365-0656 ISSN 2365-0664 (electronic)
Springer Natural Hazards
ISBN 978-3-031-66814-2 ISBN 978-3-031-66815-9 (eBook)
https://doi.org/10.1007/978-3-031-66815-9

This Springer imprint is published by the registered company Springer Nature Switzerland AG
The registered company address is: Gewerbestrasse 11, 6330 Cham, Switzerland

Contents

"You Can Learn from Bad Experience"—Reasons and Background for Palaeohydrological Investigations

1

Floods are among the most frequent and consequential natural disasters on Earth. They occur in all environments at all times. They frequently surprise with unexpected magnitudes and the damage they cause. Characteristics such as floods of the century and millennium including comparisons with the biblical deluge are quickly on everyone's lips, but what do we really know about the floods of the past? How to classify exceptional flood events like that of the River Elbe in 2002 (Fig. 1.1), the exceptional widespread monsoon rain in Pakistan in 2010 leaving vast areas inundated for months Scott (2011), or in the city of Derna located in Libya on an alluvial fan where protective dams busted after unusual intensive rain in September 2023 (Rafferty 2023)?

Have floods of this kind occurred before, are they a novelty, or are they even consequences of ongoing climate change? To answer these questions, only a look into the past can help. But typically, the records of river gauges have a limited duration, only in exceptional cases going back more than a hundred years. For large regions on Earth there are no measured values at all, but floods have left their marks here too. This may be the case in the form of records and descriptions from historical times, which go back differently far in different cultural circles. It is therefore not surprising that there is practically no information on floods in historical times from e.g. the western USA, because the indigenous population did not practice any written records before the European appeared on the continent. The situation is different for the early advanced civilisations like in Ancient Egypt, China or Mesopotamia or the regions from which a long period of historical writing is available. However, the example of the historical flood levels of the River Tiber in Rome also illustrates that over longer periods of time, the information can become less precise. For example, even though the report of flood events here dates back to the fourth century B.C., information on water levels is only available from the twelfth century A.D. (Fig. 1.2).

In addition to the historical records, natural indicators in the form of deposited sediments or traces of erosion may have been preserved and thus provide information about the water level of floods. This is particularly important for the prehistoric period, but also currently for regions without gauges. This quickly leads to further questions, such as for maximum floods (Fig. 1.3) or the oldest observations ever.

When comparing the discharges of different flood events from historical and prehistoric times (cf. Chap. 3 for references), it quickly becomes clear that the extreme floods in Central Europe, namely of the recent River Rhine with peak values of 11,000 m^3/s, rank at the lower end of the scale. It also becomes obvious that different causes of flooding can lead to discharges of much higher magnitudes. But how do flood discharges that can exceed 1,000,000 m^3/s actually develop? And how do we actually obtain data of floods that no one has observed and measured directly and that occurred several millennia in the past?

1.1 What's in This Book?

Overall, the book is intended as a reader to arouse interest in the topic of flooding, which is more complex than is generally thought, especially from a geoscientific perspective. Numerous pictures and illustrations have been deliberately included to achieve the best possible clarity and to make reading the book easy and relaxed. Any reader has an individual background e.g. from geography, geology, hydrology, history or engineering sciences and might welcome some easy-to-read introduction in chapters focusing on less familiar topics and disciplines. Numerous selected references to further reading literature have also been taken into account, so the reader can easily find more in-depth information in addition to the aimed overview character of this book. Note, that occasionally references for further reading respectively a more fundamental introduction do

J. Herget, *Flood Reconstruction*, Springer Natural Hazards, https://doi.org/10.1007/978-3-031-66815-9_1

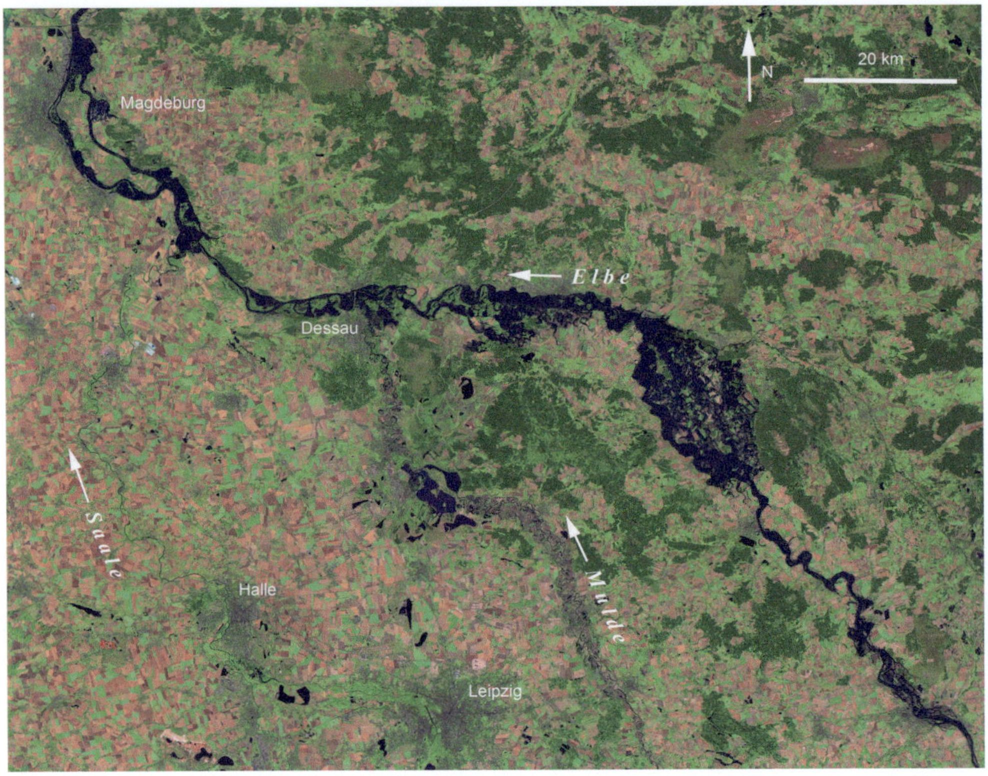

Fig. 1.1 Flooding of the River Elbe in August 2002 (NASA Earth Observatory). The barely recognisable River Saale of comparable size illustrates the magnitude and extent of the flooding in the Rivers Elbe and Mulde

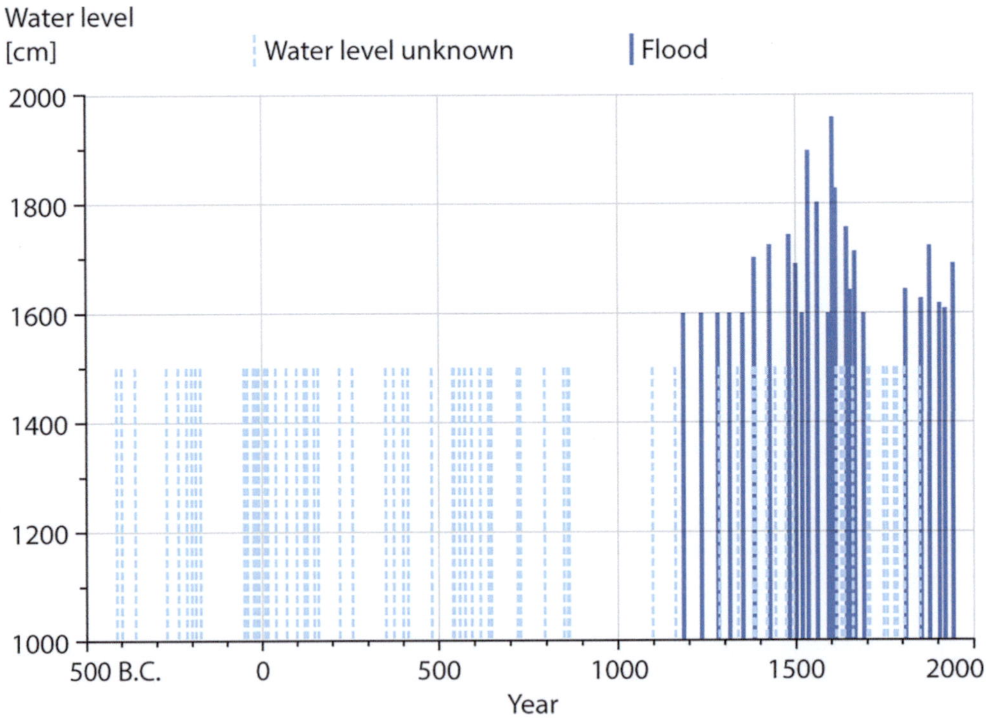

Fig. 1.2 Flooding of the River Tiber in Rome (data from Camuffo and Enzi 1996)

not suggest the most recent or comprehensive textbook, but manuals of a previous generation of publications. The main reason is a review and introductory style of the publication focusing on beginners on the topic without experiences of the discipline at all. To a minor account, also reviews with a broader background and extended reference list are considered to be used as a kind of bibliography on the topic.

The topic addressed in this book is the reconstruction of floods from historical and prehistoric times. This reconstruction focuses on central parameters such as flow

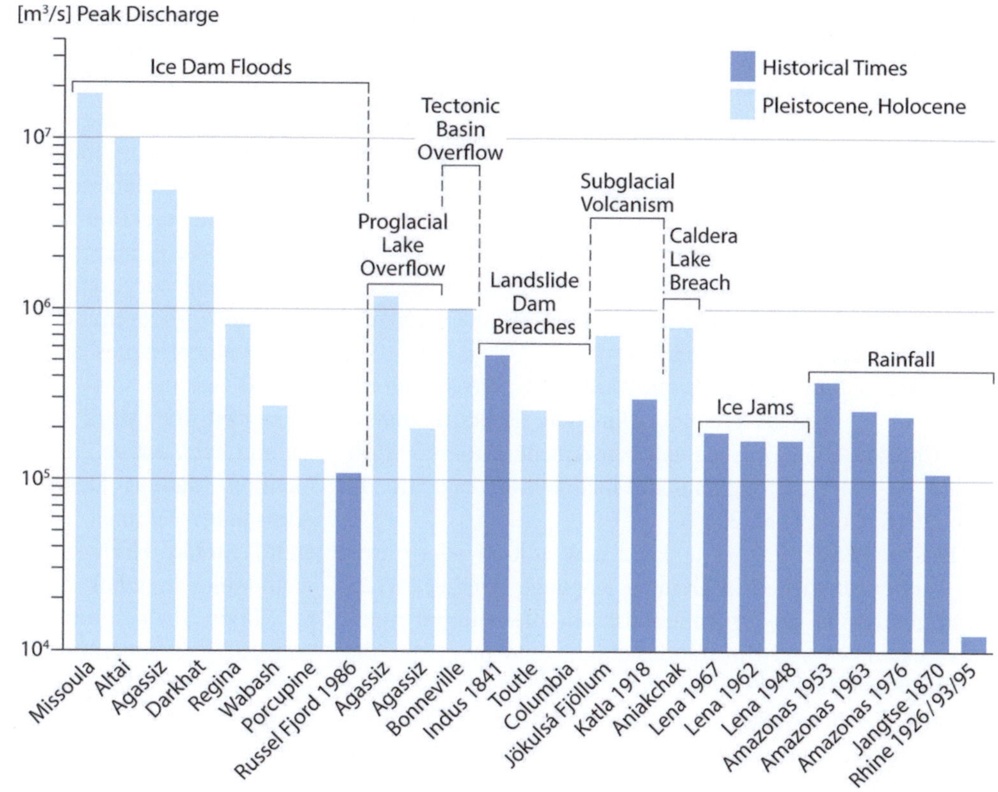

Fig. 1.3 Extreme floods of different causes since the last ice age 20,000 years ago (supplemented and modified after O'Connor et al. 2002). Note the logarithmic scaling of the peak discharge data

velocity and discharge, but also on the cause and triggering processes of flood events. In order to assure that flow velocity and discharge estimations are not coffee grounds reading, a detailed methodological section precedes the presentations on floods. It is explained in detail how water levels can be reconstructed and flow velocities estimated according to various approaches. Since quantification is involved, this part is not entirely without formulas. Nevertheless, the presentations are more of an overview and argumentations by mathematical connections, correlations and consequences are avoided. Simple, frequently empirical approaches are preferred over complex integrated software solutions for transparency and applicability for everyone.

The documentary part begins with studies of floods in historical times. These are structured regionally according to river basins and are supplemented by non-European examples. Only exemplary regions can be dealt with here, but extended tables provide references to key literature on catchments and regions not dealt with here in further detail. In addition, selected individual flood events are presented that had significance in Central Europe and left striking traces in many places, and the question of periods of increased flood frequency is addressed. This section shows parallels in parts with presentations on historical climatology (White et al. 2022). Therefore, the climatological aspects will not be dealt with further here in order to avoid repetition (cf. reviews by e.g. Pfister and Wanner 2021; Glaser 2008; Brázdil et al. 2005; Bradley and Jones 1992; Alexandre 1987). Having thus presented a broad spectrum of findings on floods, the section on historical times will conclude by demonstrating the reconstruction of flow velocity and discharge of a flood event of the River Rhine in late mediaeval times in Cologne. The time frame of the presented studies on historical climatology is supplemented by representations by floods that occurred in prehistoric times without human documentation but after the last ice age in the Holocene.

As the overview of peak discharges of floods in the past illustrates, lake outbursts have a special significance for floods of large magnitudes (Fig. 1.3). Since this cause of floods is hardly present and receives minor attention in the available flood hazard literature, it is addressed here accordingly. Not only examples of corresponding events are presented, but also the background such as formation, dynamics of the lakes as well as the processes involved in the failure of ice-dammed lakes, landslide basins or artificial dam walls are explained. For the sake of clarity, the time period is treated somewhat generously in this section and reservoir outbursts from more recent times are also discussed. These lakes form the basis for a better understanding of the major events from the Pleistocene, i.e. the last ice age. Floods whose discharges can be compared to ocean currents are dealt with here, as they involve magnitudes of millions of cubic metres per second and were able to reshape broad areas by erosion and deposition. There is also evidence that some of them had an impact on global climate. In this context, examples from Europe, Siberia and North America are presented a brief review, to which the highest peak discharges are attributed in the overview in Fig. 1.3.

In recent Earth history, land bridges formed that led to the isolation of sea basins such as the Black Sea and Mediterranean Sea. Their refilling was assumed to be associated with the biblical Flood or represents the strongest flood still detectable on Earth.

Floods in flowing waters are dealt with here and storm surges in the oceans are not addressed. Even though it might be somehow obvious and consequent, a rounding off review on historic droughts is not included (cf. e.g. Brázdil et al. 2018; Cook et al. 2015).

This book presents an updated new edition of a previous version published in German language (Herget 2012). Even though it was modified for an international audience, examples and illustrations might tend to favour the Central European region and such perspective on the topic. Thereby, the book tries to combine and partly update previous reviews and compilations with different, frequently more specific foci. In addition to further details below and references given in the different chapters of the book, the following exemplary collection might illustrate the range of previous reviews. In a special issue of the Journal of Hydrology titled "Analysis of extraordinary flood events" (Kirby et al. 1987), Baker (1987) provided in addition to several reviews on specific methods and results of regional studies in the United States of America and China one of the first reviews for palaeoflood hydrology as discipline. A trend of a closer focus on historic times is characterizing the reviews by Benito and Thorndycraft (2004), Brázdil et al. (2006), Benito et al. (2015), Wetter (2017) and Wilhelm et al. (2018) even though the perspectives of the authors range from geomorphological, climatological, historical to the emphasis of the interdisciplinary character of the approaches. Based on geomorphological and geological field evidence and the application of indirect methods for discharge estimation Baker et al. (1988) and House et al. (2002) compiled experiences from palaeohydrological studies focussing on pre-historic flood reconstructions. Both, key-studies and conceptional reviews on extreme hydrological events and megafloods were presented by Burr et al. (2009), Carling et al. (2020) and Herget and Fontana (2020). Recently, Baker et al. (2021) provided a perspective of recent and future trends in palaeohydrology in memory of Ken Gregory. The group working on Global Continental Palaeohydrology GLOCOPH compiled several books on palaeohydrology in general which contain conceptional ideas, water balance studies considering environmental changes on different continents and different climates including flood studies and applications that are still worth a closer look (Gregory et al. 1995; Branson et al. 1996; Benito et al. 1998; Gregory and Benito 2003).

Before explaining a few basic terms that are helpful for understanding the following explanations, some colleagues and collaborators should be thanked for their support in the creation of the book. Thomas Roggenkamp implemented numerous ideas and drafts of maps and illustrations and successfully mastered even difficult cases of literature procurement. Katja Kunze supported numerous technical issues and Irene Johannsen and Martin Gref carried out the final artwork. Numerous colleagues have supported the work by providing illustrations, materials and explanations. In alphabetical order, I would like to mention Mathias Deutsch, Giles Erkens, Thomas Euler, Alessandro Fontana, Daniel Garcia-Castellanos, Thomas Held, Stefan Harnischmacher and Jef Vandenberghe. I would like to thank Springer Publisher, namely the editor Simon Shah-Rohlfs for his interest in the topic and the constructive cooperation. Experience from various research projects, which have been gratefully supported financially by the German Research Foundation and the Volkswagen Foundation, could be incorporated in the text.

1.2 A Few Terms, Vocabulary and Disciplinarily Comments

To follow the text easily, it might be helpful to clarify the meaning of some basic terms. The aim here is not to go through an introductory glossary or to list sterile definitions, but to introduce a few terms that are used to describe and characterise floods and to understand the background of some applied approaches with origins in different disciplines.

Floods are defined as water levels or discharges that are above a chosen threshold value. This definition seems comparatively vague, but it is universally applicable and valid. This becomes particularly obvious when one moves away from the familiar picture of Central Europe and realises how diverse the hydrological conditions can be in different climatic zones. One example are desert areas, where there is no permanent but only episodic runoff (Fig. 1.4). A reference to the mean runoff or water level of practically zero for the definition of floods makes no sense as any discharge would be a flood. So how to identify a flood under such circumstances or does any discharge present already a flood under such circumstances?

The definition also implies that floods are defined in terms of their potential damage. It is usually irrelevant whether the high water level is caused by a backwater effect in the channel, for example, and does not necessarily also include an increased discharge. The more frequent

Fig. 1.4 Furnance Creek Wash in Death Valley, California, is notorious for its devastating floods, even though it by no means carries water every year

alternative is, that the flood is caused by such a large discharge that the channel bed becomes too small and the river inundates its banks. An equally high water level can therefore have been caused in different ways: with or without increased discharge.

Exceptional large floods are characterized as **megafloods**, if their immense magnitude exceeds a million cubic metres per second and the flood is of short duration (Baker 2020). This threshold is derived from the measuring unit of 1 Sverdrup Sv of such value and is used for ocean currents, which are flowing with lower energy and are longer lasting. Using the terms megaflood and superflood synonymously, Carling and Fan (2020) focus on the effect of the fracturing of sediments transported in such high energetic floods, frequently observed as Pleistocene ice-dammed lake outburst floods (cf. Chap. 5). The particle comminution indicated in related sediments differs megafloods from moderate events without a distinct threshold for the discharge.

Floods are usually associated with damaging disadvantage, which is often true and illustrated by the historic town of Wiang Kum Kam. As one of the capitals of the Lanna Kingdom in northern Thailand, it was abandoned after severe destruction by a flood approximately 500 years ago (Ng et al. 2014). On the other hand, it should not be forgotten that floods can also have something positive and life-giving. The early advanced civilisation of Ancient Egypt, for example, was based on the annual floods of the Nile, which irrigated vast areas of land and deposited fertile loam mobilized by soil erosion in the upper reaches.

Floods show a **hydrograph**, i.e. a change in discharge expressed as water volume per unit of time (usually m^3/s) over time (Fig. 1.5).

Starting from a base flow in perennial rivers, which is typically fed by groundwater or corresponds to the threshold value of the discharge to define a flood, the discharge begins to increase with the onset of the flood. The maximum value of the discharge is called peak discharge. It occurs before the discharge begins to decrease again. This change in discharge over time is called unsteady flow. Steady flow conditions, graphically corresponding to a parallel to the time axis, does not occur in practice because there are always at least minimal changes, but is assumed in some modelling for simplicity. Multi-peaked discharge

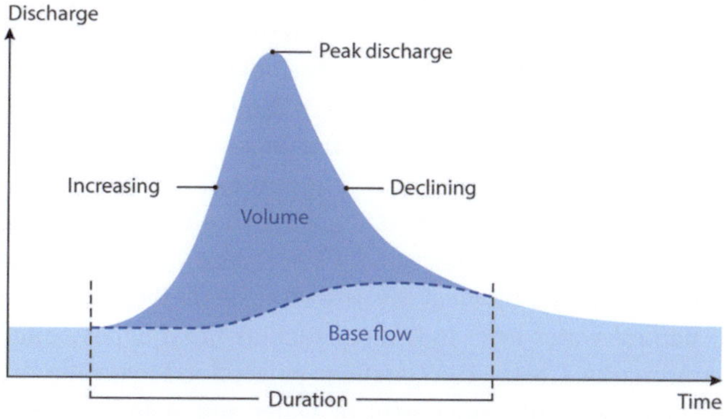

Fig. 1.5 Characteristic parameters of a flood illustrated by a schematic flood hydrograph

hydrographs with temporary increases and decreases are frequently observed. When the flood subsides, the base flow has usually increased because part of the water from the river bed, often referred to as the channel, has seeped into the subsoil. This effect and retention through wetting of surfaces as well as backwater in the floodplain of rivers mean that floods tend to decrease along their flow path, provided there are no further inflows from tributaries. This decrease is called retention loss. Since this retained water re-enters the channel with a time delay, floods often subside slowly. The rising limb of the flood hydrograph is steeper than the falling one. Exceptions confirm the rule and can be interpreted as characteristics of certain processes, for example in the case of reservoir outbursts. The volume of a flood results from the area or the integral under the hydrograph and corresponds to the sum of the discharge, e.g. the volume of a lake that drained.

There is a multitude of definitions and differentiations for the terms "**historical flood**" and "**palaeoflood**", which vary according to the discipline and the perspective of the author. In general, both terms mention floods that occurred before the beginning of systematic measurements of water levels respectively flow velocities. At a first glance, one would tend to differ the terms by the time period considered: historical and prehistoric (= palaeo) times. By a closer look, the applied terminology unfortunately is more complicated.

According to Brázdil et al. (2006, p. 741), " ..., *historical floods are events documented by indirect (noninstrumental) means of human observation and in different documentary sources.*" In addition to this, for palaeofloods the record is obtained from physical evidence in the field like flood deposits, while both are not defined by the time scale involved (ibid., p. 742). Therefore, flood events become palaeofloods if their remnants in the field are used to estimate their hydraulic parameters, independent from the point of time they occurred—which includes recent events (e.g. House et al. 2002). By references and chronologically, these definitions go back to papers by Baker (e.g. 1987) and are based on investigations carried out in USA. There, especially in the western part of the country, recent measurements and historical documentation cover a rather short period of time. A reason might be seen in short times for measurements and in missing written history by the native Americans. Therefore, field evidence of previous floods dominates documents—independent of the time period the flood occurred. Consequently, and since the focus of the hydrological investigations is on the evaluation of natural water level indicators such as flood deposits and scour marks, Costa (1987) defines palaeoflood hydrology as the science of the movement of water and sediment (!) in channels (!) before the period of hydrological records and direct measurements. Note, that no differentiation between historic and pre-historic times is made, while sediments are also considered and the water level remains inside the river channel without inundating the floodplain. Further on, Baker (2008) defines historical floods as those observed or documented by non-hydrologists (!), without differentiating them in time. In addition to this, he defines palaeofloods as past or ancient floods, and therefore excludes (!) recent event's investigation with palaeohydrological methods (ibid., p. 1). The picture becomes even more confusing when one takes a look at the working practice. Palaeohydrological methods are always applied when direct measurements cannot be carried out. The reason could be that simple that there are no gauges installed (Baker 2008) or they are destroyed by the flood of interest (e.g. Roggenkamp and Herget 2024). Consequently, approaches for the estimation of discharge or flow velocity of such flood events are called indirect methods (e.g. Rantz et al. 1982; Baker 2008), which nowadays includes computer-based modelling.

The contradictions and confusion by using the time-related prefix "palaeo" respectively the adjective "historical" for materials investigated are obvious and culminate in palaeofloods occurring today. Even though, this can be easily explained by the development of the discipline (cf. below for details), it is less convincing getting started with the topic. Consequently, an approach for a more plausible terminology might be useful. By a closer look, terms on methods (direct/indirect), evidence (direct observation/reconstructed estimation based on field evidence/reconstructed estimation based on human documents) and time periods (recent/historic/pre-historic) have to be sorted. A structured review for further discussion is presented in Table 1.1.

Floods occurred in pre-historic times are called palaeofloods and can only be investigated by field evidence. Pre-historic times end by definition with the first remaining historic documents. Therefore, the flood events occurred and documented contemporary since then might be labelled

Table 1.1 Relationship of palaeofloods, historic floods and recent floods to time periods and applied methods for analysis

Palaeo floods	Historic floods	Recent floods	
–	–	✓	Direct methods e.g. gauge data
–	✓	✓	Indirect methods Based on human documents
✓	✓	✓	Indirect methods Based on field evidence
Pre-historic times	Historic times	Recent times	

as historic floods. The period of historical times varies depending on the cultural area, so that no universal temporal delimitation can be established. In regions not inhabited by man or without written history, there might be no historic floods sensu stricto. As soon as direct measurements of flood-related hydrological parameters are (more or less) systematically carried out during the event, the flood might be characterized as recent. Note, that the recent flood period has a distinct beginning by the start of the recording of data. Historic times begin usually less sharp with some so far found first mentioning of floods while indicators of palaeofloods can be found far back in Earth's history (cf. Chap. 6).

Indirect methods for the estimation of flood related hydrological parameters can be applied for any previous event, both in historic and pre-historic times. Only recent floods can be measured directly during the event. Indirect information by field evidence to estimate parameters can be used for any time period (e.g. deposition of remaining sediments for water level elevation) while human documents are only available since historic times until today.

The concept is presented for discussion here for the first time and for sure might have additional weak points than e.g. potential confusion on how to characterize historic gauge data (e.g. from a gauge, that does not exist anymore: the data are for sure historic).

As illustrated in the context of the explanation of the terms palaeoflood and historical flood above, the **history and development of flood reconstruction** as scientific discipline is slightly more complex than one might expect. Perspectives of geoscientists and historians have to be combined for a general review, while engineers only recently have started to reconstruct floods not being listed in available data sets at all. In this context, the initiative "Prediction in Ungauged Basins PUB" of the International Association of Hydrological Sciences IAHS could not fulfil expectations arising from the title as the focus remained on the improvement of the modelling of already available data (Hrachowitz et al. 2013). Consequently, some references on previously published reviews might be sufficient in this brief introduction. The geoscientific perspective focussing on the interpretation of field evidence remaining from floods is reviewed in Baker (2008) and related benchmark papers are compiled and commented in Baker (2014, p. 289–344). The development of the contribution of historical sciences for flood reconstruction is considered in e.g. Brázdil et al. (2006) and Benito et al. (2015) while Gregory et al. (2015) conclude from a review of the development of different activities in recent decades for the benefits of interdisciplinary.

References

Baker VR (1987) Paleoflood hydrology and extraordinary flood events. J Hydrol 96:79–99

Baker VR (2008) Paleoflood hydrology - origin, progress, prospects. Geomorphology 101:1–13

Baker VR (2014) Palaeohydrology. In: International association of hydrological sciences (ed): Benchmark Papers in Hydrology 9. Wallingford (IAHS), pp 1–494

Baker VR (2020) Global megaflood paleohydrology. In: Herget J, Fontana A (eds) Palaeohydrology—traces, tracks and trails of extreme events. Cham (Springer), pp 3–28

Benito G, Brázdil R, Herget J, Machado MJ (2015) Quantitative historical hydrology in Europe. Hydrol Earth Syst Sci 19:3517–3539

Benito G, Baker VR, Gregory KJ (eds) (1998) Palaeohydrology and environmental change. Chichester et al. Wiley

Branson J, Brown AG, Gregory KJ (eds) (1996) Global continental changes—the context of palaeohydrology. Geological Society Special Publication 115. Geological Society, London

Brázdil R, Kundzewicz ZW, Benito G (2006) Historical hydrology for studying flood risk in Europe. Hydrolog Sci J 51:739–764

Camuffo D, Enzi S (1996) The analysis of two bi-millenial series—Tiber and Po River floods. In: Jones PD, Bradley RS, Jouzel J (eds) Climatic variations and forcing mechanisms of the last 2000 years. Springer, Berlin, pp 433–450

Carling PA, Fan X (2020) Particle comminution defines megaflood and superflood energetics. Earth Sci Rev 2024:103087

Costa JE (1987) History of paleoflood hydrology in the United States, 1800–1970. In: Landa ER, Ince S (eds) History of hydrology. History of Geophysics 3. American Geophysical Union, Washington, pp 49–53

Gregory K, Herget J, Benito G (2015) Hydrological extreme events in historic and prehistoric times. Z Geomorphol 59(Suppl. 3):1–13

Gregory KJ, Benito G (eds) (2003) Palaeohydrology—understanding global change. Chichester et al. Wiley

Gregory KJ, Starkel L, Baker VR (eds) (1995) Global continental palaeohydrology. Chichester et al. Wiley

Herget J (2012) Am Anfang war die Sintflut – Hochwasserkatastrophen in der Geschichte. Wissenschaftliche Buchgesellschaft/Theiss, Darmstadt

House PK, Webb RH, Baker VR, Lavish DR (eds) (2002) Ancient floods, modern hazards—principles and applications of paleoflood hydrology. American Geophysical Union, Washington

Hrachowitz M, 25 co-authors (2013) A decade of Predictions in Ungauged Basins (PUB)—a review. Hydrolog Sci J 58:1–58

Ng S, Wood SH, Ziegler AD (2014) Ancient floods, modern hazards—the Ping River, paleofloods and the 'lost city' of Wiang Kum Kam. Nat Hazards. https://doi.org/10.1007/s11069-014-1426-7

O'Connor JE, Grant GE, Costa JE (2002) The geology and geography of floods. In: House PK, Webb RH, Baker VR, Lavish DR (eds) Ancient floods, modern hazards—principles and applications of paleoflood hydrology. American Geophysical Union, Washington, pp 359–385

Rafferty JP (2023) Libya flooding of 2023. Encyclopedia Britannica, 27 Sep. 2023, https://www.britannica.com/event/Libya-flooding-of-2023 (Accessed 18 May 2024)

Rantz SE and 16 co-authors (1982) Measurement and computation of streamflow – volume 1, measurement of stage and discharge. US Geological Survey Water-Supply Paper 2175

Roggenkamp T, Herget J (2024) Flood reconstruction—the unexpected rather frequent event at River Ahr in July 2021. Global and Planetary Change https://doi.org/10.1016/j.gloplacha.2024.104541

Scott M (2011) Heavy rains and dry lands don't mix—reflections on the 2010 Pakistan Flood. NASA Earth Observatory https://earthobservatory.nasa.gov/features/PakistanFloods (Accessed 18 May 2024)

White S, Pei Q, Kleemann K, Dolák L, Huhtamaa H, Camenisch C (2022) New perspectives on historical climatology. Wires Clim Change 2023(14):e808

Further Reading

Previous reviews on palaeohydrological methods and results focussing on flood reconstruction are presented by:

Baker VR (1987) Paleoflood hydrology and extraordinary flood events. J Hydrol 96:79–99

Baker VR, 14 co-authors (2021) Fluvial palaeohydrology in the 21st century and beyond. Earth Surface Proc Landforms 47:58–81

Baker VR, Kochel RC, Patton PC (eds) (1988) Flood geomorphology. Wiley, New York

Benito G, Thorndycraft VR (eds) (2004) Systematic, palaeoflood and historical data for the improvement of flood risk estimation—methodological guidelines. Centro de Ciencias Medioambientales, Madrid

Benito G, Brázdil R, Herget J, Machado MJ (2015) Quantitative historical hydrology in Europe. Hydrol Earth Syst Sci 19:3517–3539

Brázdil R, Kundzewicz ZW, Benito G (2006) Historical hydrology for studying flood risk in Europe. Hydrolog Sci J 51:739–764

Burr DM, Carling PA, Baker VR (eds) (2009) Megaflooding on Earth and Mars. Cambridge University Press, Cambridge

Carling PA, Baker VR, Herget J (eds) (2020) Megaflooding on Earth—a global perspective. Earth-Sci Rev 210F

Herget J, Fontana A (eds) (2020) Palaeohydrology—traces, tracks and trails of extreme events. Springer, Cham

House PK, Webb RH, Baker VR, Lavish DR (eds) (2002) Ancient floods, modern hazards—principles and applications of paleoflood hydrology. American Geophysical Union, Washington

Kirby WH, Hua S-Q, Beard LR (eds) (1987) Analysis of extraordinary flood events. J Hydrol 96:3–383

Wetter O (2017) The potential of historical hydrology in Switzerland. Hydrol Earth Syst Sci 21:5781–5803

Wilhelm B, 21 co-authors (2018) Interpreting historical, botanical, and geological evidence to aid preparations for future floods. WIREs Water 2018;e1318

(Note, that this compilation should not be considered as bibliography as it is far from being complete and e.g. most conference proceedings are not considered)

Reviews on conclusions of investigations in historical climatology are presented with a varying focus on time periods, regions and contexts by:

Alexandre P (1987) Le climat en Europe au Moyen Age. Paris (École des hautes études en sciences sociales)

Bradley RS, Jones PD (eds) (1992) Climate since AD 1500. Routledge, London

Brázdil Pfister RC, Wanner H, von Storch H & Luterbacher J (2005) Historical climatology in Europe—the state of the art. Climatic Change 70:363–430

Glaser R (2008^2) Klimageschichte Mitteleuropas—1200 Wetter, Klima, Katastrophen. Wissenschaftliche Buchgesellschaft, Darmstadt

Pfister C, Wanner H (2021) Klima und Gesellschaft in Europa. Bern (Haupt)

Results and methodological challenges for the quantitative reconstruction of droughts in historic times are reviewed in:

Brázdil R, Kiss A, Luterbacher J, Nash DJ, Reznickova L (2018) Documentary data and the study of past droughts – a global state of the art. Climate of the past 14:1915–1960

Cook ER, 56 co-authors (2015) Old World megadroughts and pluvials during the Common Era. Science Advances 1–10, e1500561

Flood definitions and explanations of hydrological processes involved are available from any general textbook on hydrology

The specific characteristics of megafloods are presented in a review including several examples by:

Baker VR (2020) Global megaflood paleohydrology. In: Herget J, Fontana A (eds) Palaeohydrology - traces, tracks and trails of extreme events. Springer, Cham, pp 3–28

The development and future perspectives of the discipline and concepts for palaeo- and historical flood analysis including quantification is presented in:

Baker VR (2008) Paleoflood hydrology - origin, progress, prospects. Geomorphology 101:1–13

2 Methods of Parameter Estimation

In principle, the same parameters can be used to describe the dimensions of historical floods as for recent floods. However, as they cannot usually be recorded as precisely as for recent events, one usually has to live with estimates of magnitudes or be satisfied with incomplete representations. Nevertheless, contemporary descriptions have often survived that allow an astonishingly accurate reconstruction of the water level height. For example, the following description of the "Thuringian Flood" of May 1613 is available from Mühlhausen, located at the River Unstrut:

> [...] At St. George's Church the water stood 6 quarters of a cubit less than 2 inches high, and if it had risen another quarter of a cubit, it would have reached the baptismal font, it went in to the Gömer Gate, and washed over Kilian's alleys, and went down into peoples' cellars, [...].
> (according to Mühlhauser Bilderchronik from Deutsch and Pörtge 2002).

Similar to the example above, more or less exact information on the water level is often available, but the statements on the dynamics of the flood become increasingly blurred. Meaningful reconstructions can hardly be derived on this basis. For example, a chronicle of the Franconian town of Tauberbischofsheim describes that a flood on 29th September 1732 had risen so quickly that 5 people drowned in their homes and the water in the monastery church had reached up to the tabernacle and flooded the streets up to the parish church (according to Berberich from Weikinn 1963).

Even if individual chronologies allow amazingly far-reaching and detailed conclusions, the historical data on water levels have the problem that they cannot be easily transferred to today's conditions. This is due to river engineering measures that have changed the transverse profile of both the river channel and the floodplain. and thus the capacity of the river channel. First and foremost, these include bank obstructions through meander cuttings, dikes, bank reinforcements or groynes, but also dredging of the river channels for navigation (cf. e.g. Park 1977a, b; Kalweit 1993; Herget et al. 2005). Figure 2.1 illustrates the drastic changes in the bed of the River Emscher in the Ruhr District and Fig. 2.2 illustrates the extent of bed changes in the Upper Rhine since the corrections by Tulla in the nineteenth century.

Due to the changes in the channel bed since historical times, the same discharge today would lead to a different water level, as the cross-section has been changed. Accordingly, historical water levels say nothing about flood levels today, even if the discharge is identical (Fig. 2.3). In order to be able to use the information of magnitudes of historic floods, their discharge must first be estimated and, in a second step, it must be checked which water level this discharge would lead to today. But if the historical records generally have only limited precision, how can one reliably estimate discharges on this basis?

2.1 Considering the Discharge Beyond the Flood Level

A look at simple hydraulic principles helps to answer the question of which variables control runoff and how it can be reconstructed.

Basically, the discharge Q is formed as the product of the flow velocity v and the cross-section area A through which the flow passes: $Q = v \times A$. This relationship becomes obvious when looking at the units in the equation. Thus, discharge is expressed as the volume of water displaced per unit of time—depending on the scale, litres per second [l/s] or cubic metres ($1\ m^3 = 1000\ l$) per second [m^3/s] are common units here. A velocity is expressed as distance per unit of time, e.g. metres per second [m/s]. The cross-section area can be expressed in square metres [m^2], where the multiplication of velocity and area results in volume per time: $m/s \times m^2 = m^3/s$. Thus, in order to estimate the discharge of a flood in the past, "only" the cross-section area flowed through and the flow velocity need to be determined, since the discharge can be calculated from these.

J. Herget, *Flood Reconstruction*, Springer Natural Hazards, https://doi.org/10.1007/978-3-031-66815-9_2

Fig. 2.1 **a/b** River engineering using the Emscher (Ruhr District, Germany) for example. Undoubtedly one of the most drastic examples of river expansion is the Emscher in the Ruhr area, which had to be expanded into an open sewer in the course of industrialisation and is now being gradually reconstructed (https://www.eglv.de/en/). The picture on the left (**a**) shows a section of the Old Emscher in the city park of Oberhausen, on the right (**b**) the Emscher in Wanne-Eickel with the Emscherbruch landfill site in the background. The idyll in the city park is deceptive: in both reaches the Emscher flows squeezed between high dikes. Both are purely artificial channels whose shape and geometry do not allow any conclusions to be drawn about the course of the river in historical times

2.2 Water Level Reconstructions—Interpreting Flood Markers and Palaeostage Indicators

The cross-section area A can be reconstructed from the water levels that have been handed down, taking into account the topography in the transverse profile of a river valley (Fig. 2.4). The separation of the cross-section area into homogenous units (settlement, agricultural use, forested, …) is useful for the estimation of the varying related hydraulic roughness, which effects the flow velocity differently in each unit (cf. Sect. 2.3).

When reconstructing the cross-section area A, the condition in historical times must of course be recorded, i.e. interim fillings, excavations, embankments or changes in the river course must be taken into account. How this can be done is illustrated by an example in detail in Chap. 3.

Within the different units of the cross-section area, a variety of different water level indicators is frequently found. Historical information based on human observation including flood markers, written descriptions and archaeological indicators are typical for settled areas. Physical evidence by e.g. sedimentary deposits and changes in the vegetation cover are characteristic for floodplains outside villages. Rather challenging could become the reconstruction of the natural historical river channel depths as systematic modifications to ease navigation are frequently carried out and surveys of the original natural channel from historic times are rare. Typical characteristics and hints for the interpretation for these three different units respectively origin of flood level indicators are given in the following chapters. Note, that a focus is given on indicators that allow quantitative reconstruction of the flood level beyond indicating only the existence of a flood event in general like coarse grained layers in lake sediments or on floodplains.

2.2.1 Historical Flood Information

The information of flood levels from historical times can take place from various sources, which are compiled in a simplified overview in Table 2.1. In most historic records, describing the damage and destruction by a flood was priority. Consequently, it is not surprising that in many (up to most) historic records no information to derive the flood level from is given.

All sources require a critical analysis of their validity and objectivity, a so-called source criticism (cf. among others for a general review Hørland (2012) and specifically for historic floods/climatology Ingram et al. (1978), Alexandre (1987) and Glaser (1996)). In the case of historical records, it is important to check with what intention and by whom

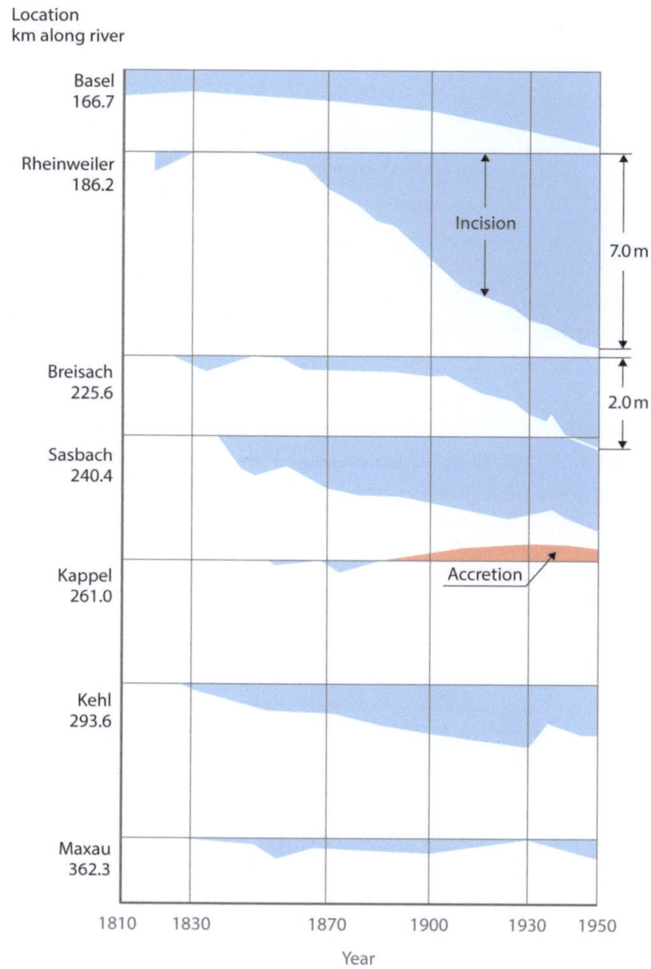

Fig. 2.2 Changes of the channel bed of the River Rhine between Basel and Maxau after the Upper Rhine Correction by Tulla and the installation of groynes and guide rails (after Kunz from Kalweit 1993). Due to the constrictions, the Rhine has incised on a significant length and thus increased the capacity of the channel bed. Therefore, flood levels of similar discharge decreased by time

the descriptions were written, since an objective description of facts cannot always be expected. If, for example, letters of request to a sovereign for reconstruction aid or tax rebates are evaluated after a destructive flood, certain exaggerations of the severity of the event and the damage that occurred cannot be ruled out. Of course, the different measurements and their conversion from historical times using as well as calendar reforms in the meantime must also be taken into account (Gyllenbok 2018). Since many data originate from the time before the first standardisations and there can be significant differences in the conversion to the metric system, this occasionally explains the different reports of water levels from different sources. The conversion of the high water level in May 1613 in Mühlhausen, Thuringia, from the contemporary source quoted above results in a water level of 81 cm in the area of the church and 14 cm below the baptismal font. Note, that a standardized scheme for checking historic information for validity is no possible as the historic information is not given in standardized structure. Concepts from historical climatological studies (cf. reference given before) might be applicable for orientation.

But even supposedly objective flood marks on buildings can cause problems. For example, many building stones tend to absorb water through the capillary suction effect within the rock even above the flood level. As a result, the rock becomes uniformly moist even above the maximum flood level and, after the water recedes, passes on an excessive maximum water level, which can be the cause of incorrect placement of the high-water mark. In some places, historical high-water marks on buildings are thankfully preserved and transferred during reconstructions and renovations, or even compiled into an overview on individual buildings from the urban area (Fig. 2.5), whereby hopefully no transfer errors occur.

Unfortunately, these problems are not always immediately obvious, as in the example in Fig. 2.6, where

Fig. 2.3 Flood levels of the River Ahr of 1804 and 2021 in the village of Dernau, Eifel Mountains, western Germany. Both events had a similar discharge while the younger flood event reached a significantly higher level due to decreased cross-section area resulting from an increased density of the settlement within the narrow valley bottom (Roggenkamp and Herget, in review)

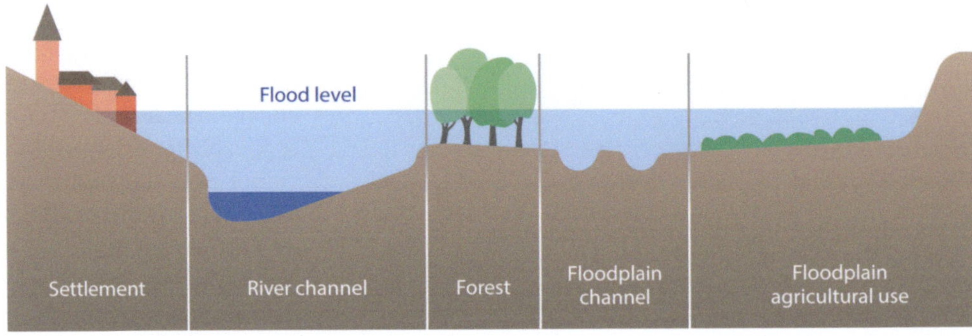

Fig. 2.4 Schematic diagram of a cross-section area of a flood level with a varying floodplain topography

Table 2.1 Historic anthropogenic sources for the analysis of floods (supplemented after Deutsch et al. 2010)

Printed sources	• Chronicles • Monography on specific events • Leaflets • Regional descriptions and histories • Journals
Handwritten sources	• Chronicles • Diaries, calendars • Official damage reports • Tax documents • Protocols of administration conferences • Gauge files and logbooks • Documents of water management authorities
Concrete sources	• Flood markers • Flood monuments • Inscriptions at mills, bridges, town gates • Archaeological evidence
Maps/plans	• Maps of inundation areas • Hydrographic maps • Position plans of dykes and dams • Construction plans of e.g. bridges, dykes, dams, …
Illustrations	• Contemporary pictures and etches of floods • Painting of rivers during floods • Townscapes and etches • Flood photographs

a building block with a flood mark was used again in the reconstruction of the railway embankment, but was installed in reverse. The flood marker for the January 25th 1374 water level in Düsseldorf-Urdenbach is affixed on a house built in the mid of the eighteenth century (Fig. 2.7). A speculative interpretation could be that the focus was on the even higher flood level in 1784, which maybe should be pronounced for being even above the legendary flood level 400 years ago at the location at the at that time recently built house—probably, no levels of the ancient flood were known beyond being the largest in peoples' memory.

It has therefore proven useful to consider several sources, both in terms of their type and origin, in order to recognize contradictory and erroneous reports and to be able to correct them if necessary. This is especially true for collections of historic texts on flood events, which should not be interpreted unchecked (cf. review table in Chap. 3).

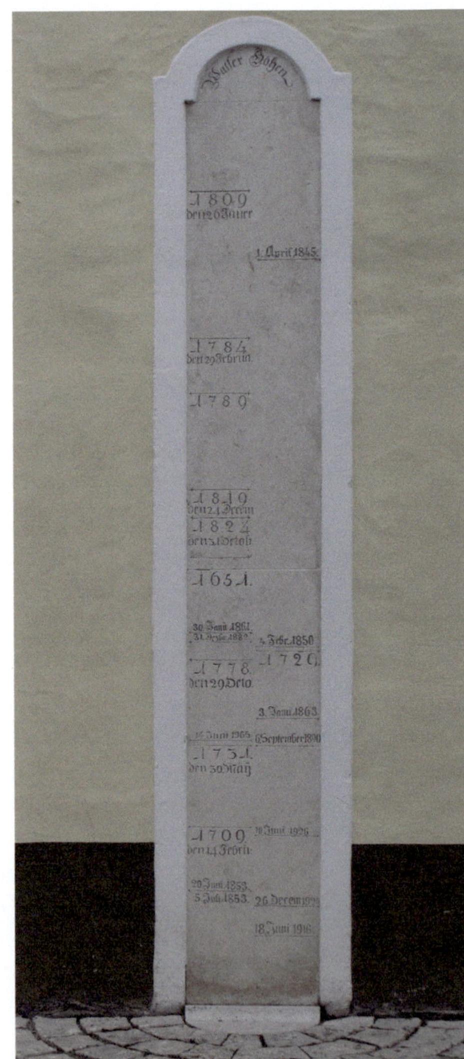

Fig. 2.5 Compilation of flood levels of the River Danube at Kelheim, southern Germany

Fig. 2.6 Building stone with flood marker for the 1906 flood of the River Rhine in the abutment of a railway bridge in Koblenz-Ehrenbreitstein, which was installed in reverse when it was reused after the destruction of the bridge

Fig. 2.7 Flood levels of the River Rhine in Düsseldorf-Urdenbach—the flood level from 1374 on a building constructed in the mid of the eighteenth century requires explanation

Plausibility checks are possible by compiling source data of a flood event along a river course at different points and relating them to each other, e.g. with regard to the increase in discharge downstream, after the confluence of larger tributaries or for the chronology.

Archaeological evidence provides additional hints on water levels in historic times. Flood levels are indicated by the location and dimension dams, dykes or even historic gauges not only in the old cultures in Mesopotamia or Ancient Egypt (e.g. Viollet 2007). By arguing, that the urban waste water system of the city of Cologne in Roman times should have been in function even during average flood levels, Roggenkamp and Herget (2016) estimated the mean flood level of the River Rhine at Cologne by the elevation of the drainage outlet into the river channel. Note, that flap valves preventing the river to enter the drainage system during flood events, were not invented those days. E.g. van der Meulen et al. (2022) identified fluvial layers among the archaeological heritage of the city of Arnhem and could relate them to the largest historic flood of the River Rhine in AD 1374.

Dimensions of bridges provide a remarkable amount of information on water levels and river channel conditions without any survey or measurement of related parameters at all (Roggenkamp and Herget 2016) (Fig. 2.8). River channels frequently meandered across the valley bottom while the age of the bridge fixes the location of the river channel at a location for the related period. The length of the bridge indicates the width of the channel. The lower end of the bridge piers indicates the elevation of the channel bottom, which in combination with the age of the bridge provides information on the incision rate of the river. In case there was navigation on the river in historic times, the draught of the boats indicates minimum water depth of the river channel most of the time, hence the mean water depth. The range for the mean water level is limited by considering that boats should be able to pass through below the bridge most of the time. Consequently, the height of the bridge indicates the uppermost mean water level for known heights of boats traveling along the river. Assuming, that the bridge should be usable also during average higher water levels, the height of the bridge pier limits the flood level before inundating the bridge. Therefore, the capacity of the river channel below the bridge deck and between the abutments indicates average or ordinary probably even mean flood magnitudes—the latter not by statistical definition, of course. In historic times, bridges were expensive to build and maintain, and therefore might be assumed to be built as small as suitable, which supports the argumentation of the hydrological interpretation of bridge dimensions. Destruction of bridges by floods or their rebuilding related with enlargements might indicate exceptional flood magnitudes or changing flood frequencies.

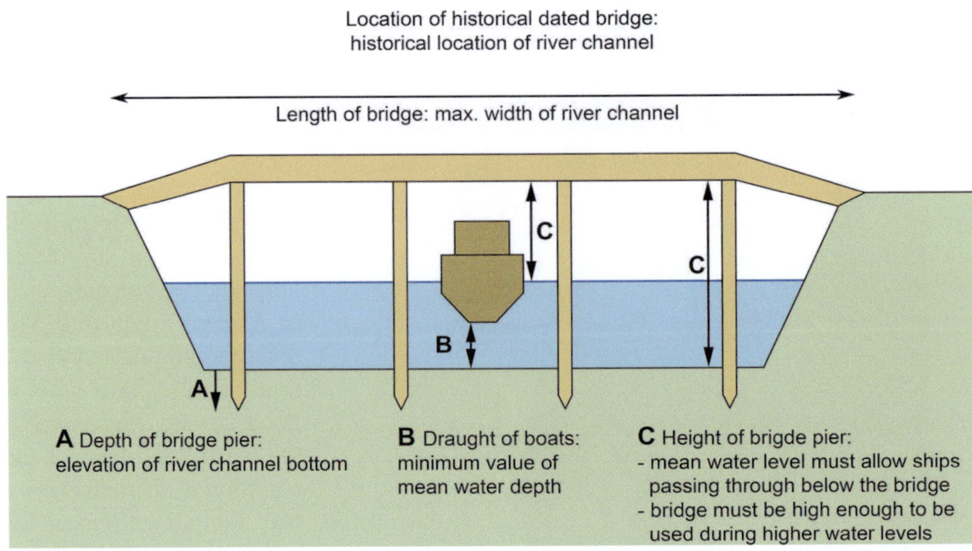

Fig. 2.8 Bridge dimensions related to water levels in rivers (modified after Roggenkamp 2023)

Fig. 2.9 Blades of grass in tree branches indicate the last flood level (flow direction left to right)

2.2.2 Natural Flood Level Indicators

Even outside populated areas and independent of written or pictorial records, water levels of past floods can be handed down. Everyone has certainly seen the trace of objects floating on the water surface after a flood. Such kind of so-called flotsam ranges from blades of grass to plastic waste, which can be found on meadows and in bushes in river floodplains (Fig. 2.9).

Such indicators of previous water levels are generally referred to as flood marker for recent and palaeostage indicators for older up to prehistoric events, whereby the time span since the flood can range from hours to centuries and millennia. Under favourable circumstances, such palaeostage indicators can also be preserved from floods of Pleistocene and Tertiary times, as examples will show in the following chapters. Figure 2.10 illustrates different types of natural flood level indicators.

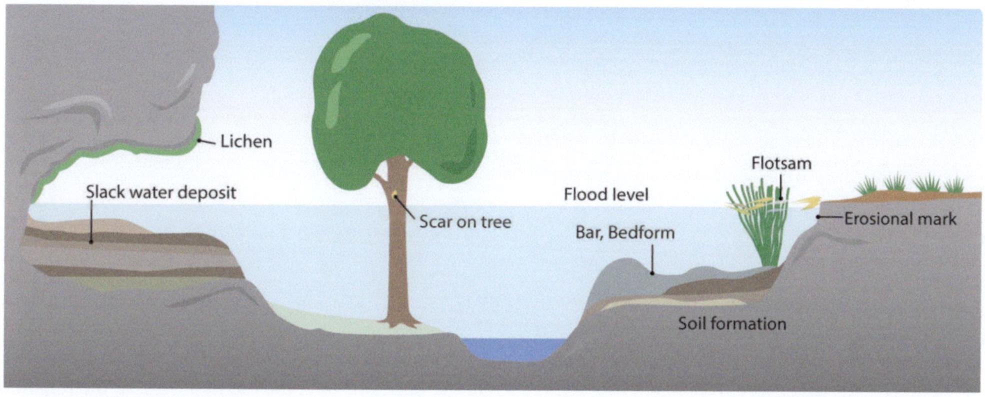

Fig. 2.10 Schematic illustration of natural flood level indicators (modified after various sources from Herget 2020)

Some traces indicate a minimum flood level, others a maximum flood level, were formed by erosion or accumulation (Figs. 2.8 and 2.9) and are related to a single flood event or characteristic recurrent flood levels. Another classification component are botanical flood level indicators as will be explained below. Anyhow, all water level indicators require critical review in terms of the precision and therefore it is recommended to avoid interpreting a single indicator in isolation.

- **Erosional flood level indicators**
  Erosional edges on riverbanks are widespread flood level indicators and can be found in different dimensions (Figs. 2.11 and 2.12). As long as the slopes do not blur these marks by sliding or they become indistinct due to vegetation cover, they are relatively easy to detect (Fig. 2.13).

Erosional edges typically indicate frequently recurring flood levels at which the edge is successively formed on the bank. In individual cases, however, high-energy single events can also form erosion edges. On undercut banks, the upper edge of post-flood slope sediments should not be misinterpreted as a high-water mark.

Repeated erosion in the area of frequent flood cover along river banks also impedes soil formation, so that the degree of soil development and soil parameters can also be used as water level indicator (e.g. Smith and Boardman 1989). Illustrative examples are shown in Fig. 2.12 with a completely eroded soil cover and Fig. 2.13 with reduced soil development indicated by the lesser vegetation cover. Deposits of soil material previously eroded in the headwaters can also accumulatively influence soil formation as colluvium.

- **Accumulative flood level indicators**
  Slackwater deposits form during floods from the transported suspension load in areas of locally reduced flow velocity, or more precisely, reduced drag force (Kochel and Baker 1988). Their formation is particularly favoured in the mouths of tributary valleys and are located in straight, moderately inclined valley sections. By the elevation of their surface, however, they only trace the minimum water level, since deposits from entrained suspended load do not necessarily accumulate to the water surface elevation during sedimentation. In individual case studies, an underestimation of the actual maximum flood level compared to the surface elevation of the stillwater deposits by 10% or in another case by 50–90 cm has been observed, resulting in an underestimation of the peak discharge by 20% (Greenbaum et al. 2000). Depending on the size of the deposition area, typical deposition sites are niches in the bedrock, areas in the flow shadow behind obstacles in the floodplain (Fig. 2.14) or local valley widenings, even the mouths of water carrying tributaries. Having originated from the entrained suspension load, deposits typically consist of horizontally stratified fine sands and coarse silts, similar to the typical floodplain loam. In several areas without any or rather short record of floods, slackwater deposits could be used to identify flood magnitudes and improve data of frequencies of rare high magnitude events (Lam et al. 2017; Wiman et al. 2023). During high-energetic megafloods significantly coarser particles can also be transported as suspension load (cf. Pleistocene outburst floods in Chap. 5).

Gravel bars (Fig. 2.15) form from the entrained bed load when the transport force decreases locally. The deposition is caused by the reduction of drag force and flow velocity as a result of a local widening of the river course. Indeed, the continuity formula given above illustrates that if the discharge remains constant, the flow velocity v must decrease as the cross-section area A increases. Hydraulic details on shape development and differentiation of gravel bars according to flow conditions are given by Carling (1989). Gravel bars can also form in local current shadow locations

Fig. 2.11 Accumulative and erosive water level indicators in a stream channel, Valley of the Gods, south-western Utah, USA. Looking against the direction of flow, an accumulated layer of fine sediments can be seen on the left hand side in the current shadow of the bush. On the opposite side, the erosion edge of the remaining fine sediments above the gravelly streambed also indicates the minimum water level of the last flood

Fig. 2.12 Different generations and sizes of flood level indicators in the lower Katun River, south of Chemal Village, Altai Mountains, S-Siberia (view against flow direction). While the lighter-coloured fringe immediately above the current water surface (marked by arrow) indicates the regular recent fluctuation range of the water level, the exposed rock surfaces on both sides of the river channel indicate the high water level during the annual snowmelt, which did not reach the extensive flat valley floor but eroded the entire soil cover below. The partially forested gravel bar on the left flank of the valley was deposited by Pleistocene outburst floods (details in Chap. 5) and represents an accumulative palaeostage indicator

similar to stillwater deposits growing downstream. In contrast, gravel bars formed by backwater in front of local obstacles grow upstream against the gradient and the direction of flow, as the bed load is deposited earlier and earlier. Regardless of the development of the gravel bars in detail, they have in common that they cannot grow up to the water surface and therefore indicate a minimum water level. Due to the lack of vegetation cover, they are easy to recognise and have a good preservation potential due to their comparatively coarse deposits. In case of erosion of the bar surface during the final stage of a flood event, the original height can be reconstructed based on the internal structure of the bar deposits (Almeida et al. 2016).

Based on hydraulics of boundary shear stress and its related exponential rise from the bottom, the grain size of sediments of the gravel bar transported as bed load by a flood is related to the depth of flow. Avoiding hydraulic equations and explaining the physical context here, reference should be given to an empirical relationship by Knox (1987, p. 161), who also discussed alternative approaches. Flow depth D [m] is found as $D = 0.0001A^{1.21}S^{-0.57}$ with A for the intermediate axis in mm of the five to ten coarsest particles moved and S for the energy slope in m/m (frequently parallel to the slope of the water level surface). By a closer look, several boundary conditions for the current pattern and sedimentary parameters must be considered but for undisturbed conditions this equation is valid for a grain size range of 50–3300 mm and statistically significant. The availability of particles exactly related to the depth of flow limits the application of the approach, so in practise minimum values for the depth of flow might be indicated by the coarsest grain sizes. At least for plausibility tests of other flood level indicators this approach might provide a suitable verification.

Note, that bedforms like fluvial dunes and obstacles marks developed in the river channel or on the floodplain are also related to the depth of flow during their formation. To ease the review, these aspects are mentioned in context with their indication of flow velocity in the following chapter.

Fig. 2.13 Erosional flood level indicators in Palouse River valley, Washington, USA (flow direction to the top right). The hillslope debris is eroded at its lower part by high magnitude floods of the river. Note in addition the darker colour of the rock in an area immediately above the water level, which can be related to a lack of weathering crust respectively vegetation cover due to more frequent low level floods

Fig. 2.14 Slackwater deposits consisting of silts and fine sands at French Creek, South Dakota, USA

During floods, large amounts of plant debris but also human waste are transported floating on the water surface and deposited as a fringe along the maximum water level when the water level drops. In the introduction, reference was made to the deposition of such flotsam in bushes and trees in flood beds as a widespread water level indicator (Fig. 2.9). On closer inspection of this specific example, however, these supposedly clear indicators are not without pitfalls: the fine branch tips submerged below the flood level have caught entrained blades of grass, but have been further lowered by the flow pressure of the water through their growth against the direction of flow. This effect may have been enhanced by initial accumulations of flotsam on the branches. After the flood waters receded, the pressure of the incoming water was absent and the branch was able to spring back to its higher initial position, provided the load of the floating debris did not reduce this. This springing back and up in the absence of inflow pressure therefore potentially overestimates the height of the water level. However, time also has an effect here, since when the blades of grass hanging on the branch dry out, their weight decreases and the branch can be bent up further. Here, comparisons with the height of the flotsam or other flood markers in the immediate vicinity are helpful. In this specific case, analogous deposits were also found in nearby conifers.

An additional disturbance might result from the local increase in the water level immediately in front of obstacles around which the water flows, such as tree trunks, bridge

Fig. 2.15 Gravel bar in the floodplain of Katun River near Multa Village, Altai Mountains, Siberia (flow direction to the left). During floods, the gravel bank formed directly next to the river channel, which was subsequently eroded by cross currents and whose gravels were shifted to the adjacent sand bed

Fig. 2.16 Lichen growth on a rock surface only above the flood level, example from Canyon de Chelly, Arizona, USA. The dark horizontal band on the right rock face marks the level of the eroded soil surface on the valley floor (cf. background). Above this line, the rock face is free of lichen growth, which only starts again above the water level of the flood marked by the arrow

piers or boulders. This can be explained by the abrupt reduction of the flow velocity immediately in front of the obstacle and the associated conversion of kinetic energy into potential energy. Flotsam floating on the water surface thus reaches an elevated position in front of the obstacle compared to the undisturbed water level of the surroundings and the flood level is overestimated. However, this phenomenon can also be used to estimate flow velocity, indicated by the amount of the local water level rise (cf. explanations on run-up sediments in the following chapter).

- **Botanical flood levels indicators**
  Botanical evidence of floods occurs in a broad range of evidence: while limited lichen coverage and scars in tree rings mark distinct flood levels, the distribution of different species of vegetation within the floodplain indicate frequency, magnitude and duration of repeated inundations.

Lichens form the pioneer growth on rock surfaces and show a clear correlation with the frequency and elevation of flooding (e.g. Foulds et al. 2014). For example, Gregory (1976) was able to show that annually flooded bank areas are not colonised by lichens and that in the higher sections the lichen density increases in inverse proportion to the frequency of floods. Note, that lichens are eroded by floods. Especially in areas without gauges, estimations of flood frequencies and magnitudes can thus be found (Fig. 2.16).

Flotsam of e.g. fallen tree trunks swimming on the water surface can leave impact marks on trees in the floodplain that mark the current water surface relatively precisely and can remain as an anomaly in tree rings for a long time (Fig. 2.17). The scar indicates both the elevation of the water level and the year the event occurred by the annual resolution of the tree rings. Flood chronologies from overgrown impact marks in tree rings for regions that are otherwise poor in flood records like parts of North America (Yanosky and Jarrett 2002) have been developed. Gottesfeld (1996) was able to observe a very good correlation between recent impact marks on trees and the maximum flood levels, but also found significant over- and underestimations. The overestimations are explained by local water level rises due to strong turbulence or backwater, while impact marks below the maximum water level are attributed to collisions of flotsam before or after the passage of the flood peak. This uncertainty cannot be finally clarified according to results from comparative studies (e.g. Comiti et al. 2016; Ruiz-Villanueva et al. 2019), but the maximum height of the impact marks correlates statistically well with the magnitude of the floods (Ballesteros et al. 2011). Impacts during the waning stage of a flood are relatively rare as the flood plain is already cleaned from flotsam by the flood wave before.

There is a broad variety of interaction among the vegetation and water levels along rivers and floodplains. Basic but fundamental aspects are e.g. that aquatic plants indicate minimum water levels that are usually not fallen below. Bare banks of fluvial sediments indicate frequently reached flood levels with currents sufficient to mobilize the sediments and therefore hinder plants to root. In this elevation, morphodynamics dominate over hydrodynamics unlike in the upper elevations where perennial and woody plants are growing. A review on the interactions present e.g. Hupp and Osterkamp (1996) and Müller (1995) with a focus on hydrology and flood geomorphology rather than from a biological perspective. As result, there is specific pattern of the kind of vegetation changing from reeds and grass via bushes towards more or less flood tolerant trees (Fig. 2.18). The kind of plants and especially the species differ by vegetation zones within the regional floristic kingdoms, but general patterns develop also beyond the one presented from Central Europe, e.g. in North America (Hupp and Osterkamp 1996), Australia (Shilpakar et al. 2021) or India (Tabasum et al. 2009).

In the temperature climate zone, the main influence on the vegetation changes from river dynamics towards depth to the water table within the lower parts of the softwood floodplain section. Surveying the spectrum of plant species

Fig. 2.17 Scar of flotsam from a flood in AD 1853–54 indicated by tree rings (modified from Miller 1960)

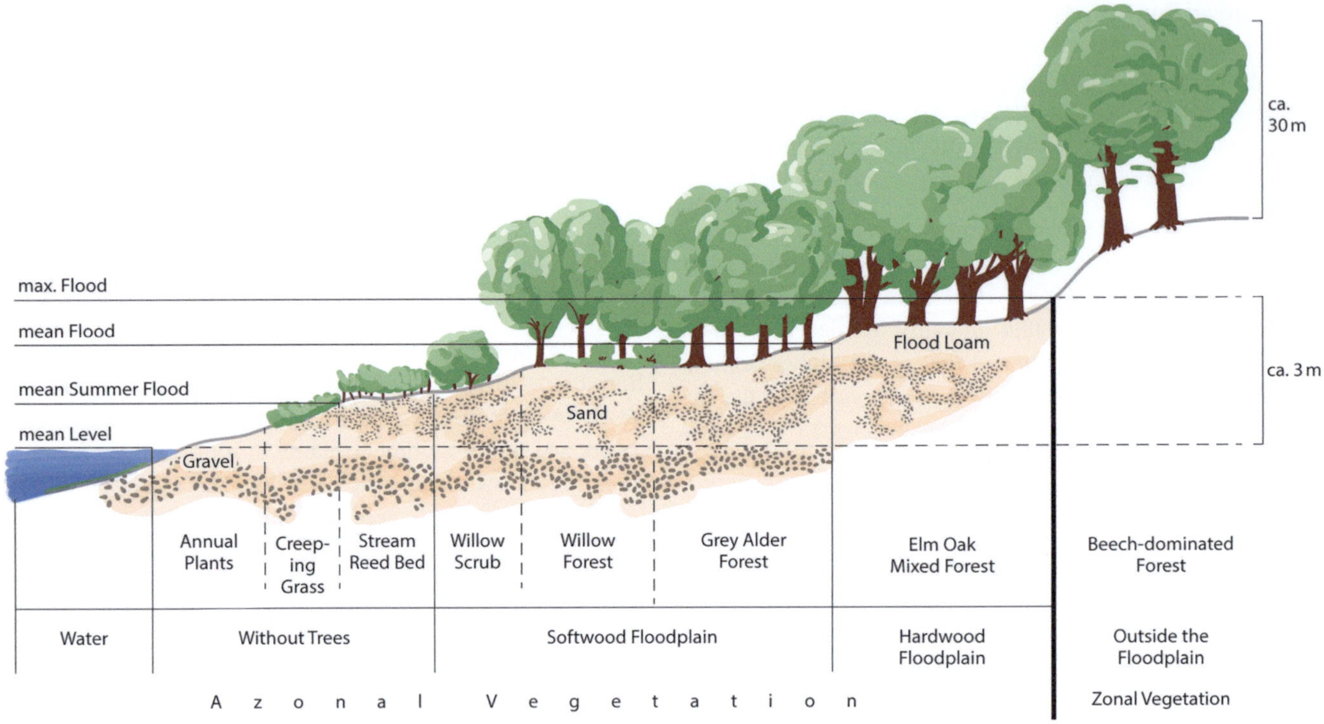

Fig. 2.18 Schematic cross-section through the sequence of floodplain vegetation in the middle reaches of a river in the foothills of the Alps, Central Europe (modified from Leuschner and Ellenberg 2017)

is challenged by the permanent succession of plants as each flood event might let the cycle start again. Analysis of the spectrum of species can be related to flood magnitudes (by the elevation of vegetation unit above mean water level), frequencies (by the specific tolerance of individual species to repeated inundation) and durations (also species-specific) (Ochs et al. 2020). For the floodplain of the River Rhine around Karlsruhe, characteristic values for inundation depth and duration are determined (Table 2.2).

Challenging might be the search for floodplain areas where the vegetation cover is still more or less close to the potential natural vegetation. For regions where even the entire floodplain is systematically modified respectively used for agriculture, still plant of specific flood related characteristic can be used as indicator. A trivial, but obvious example is the distribution of a dense cover by stinging nettles *[genus Urtica]* which indicates the extension of recent inundations by water rich in nutrients.

- **Conclusion**
  The different natural flood level indicators mentioned above can be summarized in a review table (Table 2.3) with regard to their essential characteristics. As the allocations are not always clear-cut as mentioned briefly above, make sure about understanding the related formative processes for proper interpretation.

The numerous flood level indicators along the course of large rivers are spatially scattered and—as they are related to different flood events—at significantly different elevations, which makes the overall overview more difficult. Flood traces that are comparatively easy to correlate are found along smaller rivers and streams, where the overview is easier and correlations or causes for the formation of the individual structures are easier to grasp in their context. An example of the traces of a recent flood in a small forest stream in Fig. 2.19 illustrates the context of the different flood markers.

The water level indicators in Fig. 2.19 can be interpreted as follows: During the flood, flotsam in the form of branches and twigs was carried on the water surface, blocked up in front of a tree growing in the floodplain (a). The narrowing of the culvert was further intensified by subsequent flotsam, resulting in a backwater of the runoff. The backwater increased the cross-section area, reducing the flow velocity and the related drag force. The entrained gravelly bottom load was deposited as gravel bar in the channel bed above (b). It might be expected that the first deposited gravel in turn had a backwater effect on the following bottom load, so that the gravel bank expanded upstream by growing against the flow direction. When the water level dropped, the gravel bank was divided into several units by the stream channel. A minimum value for the maximum

Table 2.2 Inundation depth and duration for the different sections of the floodplain of the River Rhine around Karlsruhe during the vegetation period (1.4.–30.9.) (modified from Michiels and Aldinger 2002)

Floodplain vegetation section	Inundation depth [cm]	Duration of inundation [days]	
		maximum	mean
Lower softwood	>270	>140	>60
Upper softwood (up to lower hardwood)	220–270	110–140	33–60
Lower hardwood	170–220	65–110	15–33
Central hardwood	90–170	35–65	4–15
Upper hardwood	30–90	10–35	1–4
Uppermost hardwood	<30	<10	<1

water level is also indicated by the slack water deposits on the far bank in the flow shadow of the tree (c, left). These could only be deposited there if they have been transported as suspended load, which gives an indication of the transport competence during the flood (cf. following chapter). The slack water sediments document a minimum value of the maximum flood level, as deposited suspended load does not accumulate up to the water surface. The elevation of the wood accumulated in front of the log also only gives a minimum value of the flood level, as the swimming wood experienced buoyancy during the flood, which subsequently subsided and may have led to a collapse of the wooden flotsam pile. However, the amounts for underestimation of the maximum water level probably might expected to be only a few centimetres in the small stream.

In studies of the frequency and magnitude of floods resulting from storms crossing the British Isles, Carling and Grodek (1994) were able to derive a typical pattern of the spatial arrangement of various flood level indicators. Their findings are probably not universally transferable literally, but offer an idea of the context of the different indicators listed above:

Table 2.3 Characteristics of different flood level indicators (simplified after Herget 2020)

Flood level indicator	Formation		Water level		Relationship	
	Erosive	Accumulative	Minimal	Maximal	Distinct event/stage	Repeated water level
Lichen	X			X		X
Tree scar	X		X	(X)	X	
Erosional edge	X		X		(X)	X
Soil	X	X	X	(X)	(X)	X
Vegetation				X		X
Slack water deposit		X	X		X	
Gravel bar		X	X		X	
Flotsam		X		X	X	

Fig. 2.19 Different fresh flood level markers along on a small forest stream in the Eifel Mountains, Germany (flow direction to the left) (Herget 2020). **a** Flotsam, backed up in front of tree trunk, **b** subsequently dissected gravel bank of sediments transported on the channel bottom, **c** slack water deposit in the floodplain transported in suspension

1. In the lowest section up to approx. 1 m above the valley floor, the surface of the solid rock is rounded by flowing water or, in the case of karstic rock, characterised by solution forms. The complete absence of vegetation including lichens indicates the annual flooding.
2. The next higher section shows fresh, bare gravel deposits with a relatively flat surface and smoothed bedrock surfaces. An only sparse lichen cover proves the repeated flooding, which is also indicated by impact marks and deposited flotsam.
3. In the upper sections, finer-grained deposits are also found, which lie on less smoothed rock surfaces. Gravelly deposits show a certain patina, which indicates that weathering of the surfaces has set in and thus that the deposit is older. At this elevation, the banks are already covered by well-developed soils.
4. The highest sections of the banks, reached only by extreme flood levels, show only minor traces of flowing water, for example by rounded gravels located in isolated places. Otherwise, the banks merge continuously into the adjacent slopes with unrounded large sediments, more developed soils and uniform vegetation cover, without showing any traces of flood influence.

In general, it can be stated that only a spectrum of multiple water level indicators can provide a sufficient and reliable indication of a flood level. When combining minimum and maximum level indicators along a river section, it is frequently possible to record accuracies of the flood level in a range of centimetres. The possibility of remobilisation of older deposits—for example rounded gravels from a higher river terrace—must be critically examined. Figure 2.20 illustrates that flood level indicators might also be found at unexpected elevations, which at a first glance are well above the indicators of current maximum flood levels.

2.2.3 Depth of the River Channel

Estimating the natural respectively historical depth of a river channel is challenging as typically surveys are rare. In the context of channel modifications to ease navigation during low water level conditions, systematic bathymetric surveys were carried out to locate and quantify required dredging volumes from the river channel bottoms. Publications on historic navigation and engineering measures in this context provide further information. Archaeological evidence of channel depth like the length and construction of bridge piers was mentioned before (Fig. 2.8). Minimum values considering the draught of boats traveling on the river might fill temporal gaps for periods without alternative information sources and become of specific interest if problems for navigation due to insufficient depth of water are reported.

Frequently, channel depth is given by single points or cross-section profiles only which require transfer to the location of flood level indicators or extrapolation. E.g. Mervalde (2009) tested a variety of spatial interpolation

Fig. 2.20 Location of an unexpected flood level indicator in the Anapodaris Gorge, Crete (flow direction to the left). Thus, the fine-grained sediment cover on the rock surfaces indicates the typical water level in the river course, which periodically falls dry. The pieces of driftwood deposited in the current shadow behind the boulder and the fresh branches in front of the boulder mark minimum water levels during floods. At the marked spot, plastic rubbish was found jammed in a crevice in the boulder, indicating a clearly higher water level of current floods. Flood level of such magnitude might even reach and relocate older river terraces (Macklin et al. 2010)

Fig. 2.21 Idealised distribution of flow velocity in a channel

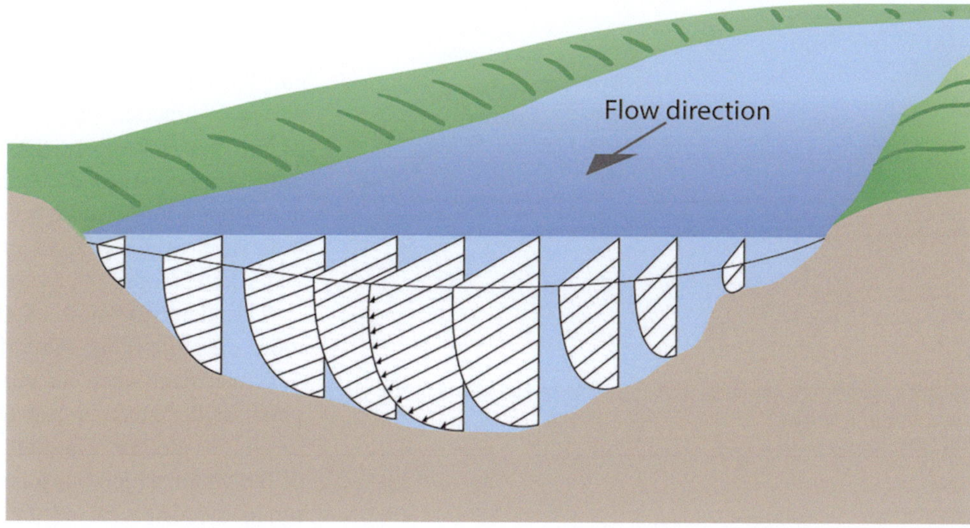

methods for the extrapolation of river bathymetry on a recent channel while van der Meulen (2021a, b) interpolated historic cross-sections towards a digital elevation model for historic times for advanced flood modelling (Bomers et al. 2019). Anyhow, uncertainties about channel depth and representative values will remain as e.g. mobile bed forms like fluvial dunes with heights of 2 m modifying river bathymetry continuously but must not lead into desperation. Considering the geometry and dimensions of the components of the cross-section area during high flood levels (Fig. 2.4) might put problems with data accuracy for the river channel itself into perspective. Variations in channel depths for channels being relatively narrow in relation to the floodplain width, result in ranges for the entire cross-section area of insignificant difference (cf. Sect. 2.5 for hints in this context).

2.3 Estimation of Flow Velocity

Compared to the rather obvious indicators of the water level, the estimation of the flow velocity during a past flood is abstract in retrospect. Following below are various approaches based on empirical approaches taking into account the main relevant influencing factors presented. Additionally, also methods that interpret flood traces and deposits hydraulically are illustrated.

Careful attention must be paid to whether flow velocities are determined as mean value for an entire transverse profile or as punctual velocities at the channel bed, bar surface or in the floodplain, which must subsequently be extrapolated to the overall profile or to mean flow velocity in a second step. Due to the decreasing influence of roughness at the channel bed and bank surfaces, the flow velocity increases towards the centre and surface of the channel bed (Fig. 2.21). This phenomenon puts the meaning of the, or to be more precise "a mean flow velocity" into perspective. Anyhow, for the discharge estimation by the equation of continuity (cf. above) only one representative flow velocity is requested and must be determined.

2.3.1 Empirical Formulas for Mean Flow Velocity

Since the eighteenth century, empirical formulas have been developed to estimate the mean flow velocity of a river without direct measurements. In principle, these formulas are also suitable for quantifying past discharges, as long as the input parameters can still be determined or reconstructed. In the essential features, the developed formulas are relatively similar, whereby the formula attributed to the Irish engineer Robert Manning (1816–1897) has found the widest distribution and most frequent application internationally. In terms of scientific history, Manning alone does not deserve the credit for naming the established approach, but his name is so common in the literature that this simplification will also be followed here. This is also done against the background that there is no agreement as to which names should be given historically justified consideration (Yen 1992).

Consciously disregarding the history of research, the corresponding correlations, further developments and derivations (Chow 1959; Yen 1992), the approach according to Manning can be explained by the fact that for practical application questions, the most important factors have been compiled in a simple empirical formula for calculating the mean flow velocity. Influencing factors relevant to practice are the shape of the channel bed, the slope and the roughness of the channel bed, from which the mean flow velocity v [in m/s] can be calculated according to the formula

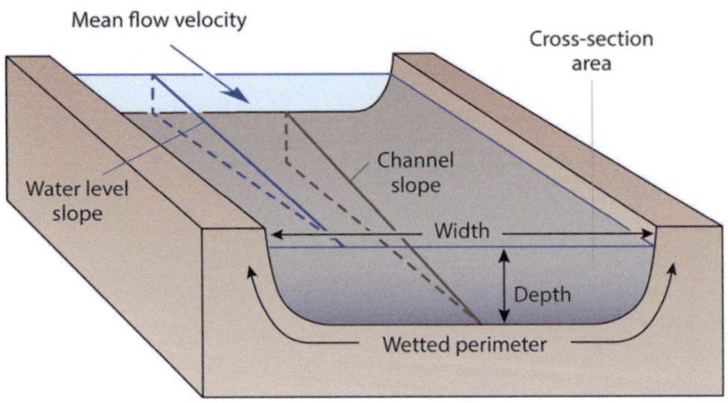

Fig. 2.22 Cross-section area, wetted perimeter and channel or water level slope of a channel

$$v = R^{2/3} S^{1/2} n^{-1}$$

The individual factors can be characterised as follows (Fig. 2.22)

The influence of the shape of the river channel is considered by the hydraulic radius **R.** This is calculated as the quotient of the cross-section area A [in m^2] and the wetted perimeter P [in m] as R = A/P and accordingly has the unit of a linear measure [m]. The values of A and P are measured in the field or can be derived from maps or digital elevation models of sufficient spatial resolution and reliability. The significance of the hydraulic radius for the flow velocity follows from the different influence of the velocity-reducing contact between the moving water body and the channel bed (Fig. 2.21), as illustrated by an example (Fig. 2.23). Thus, the flow-favourable square transverse profile results in a larger hydraulic radius R for the same cross-section area A than the flat and wide profile with a larger wetted perimeter P for the same area and thus also a higher flow velocity v.

The slope **S** is expressed by the height difference Δh [in m] per length section Δl [in m] as S = Δh/Δl. By shortening the two length measures, the slope is dimensionless or has the descriptive unit [m/m]. To ease reading slope values with typically several zeros in the decimals, frequently slopes are given in percent or even per mille—make sure not to carry out calculations with such modified values. Originally, the slope was related to the energy line gradient, the change in the energy content of the flowing water, which in practice can usually be equated with the gradient of the water level or the channel bottom for sufficiently long and undisturbed channel sections. Both slopes can be determined in the field, whereby corresponding palaeostage indicators or flood level markers can also be used for the water level slope.

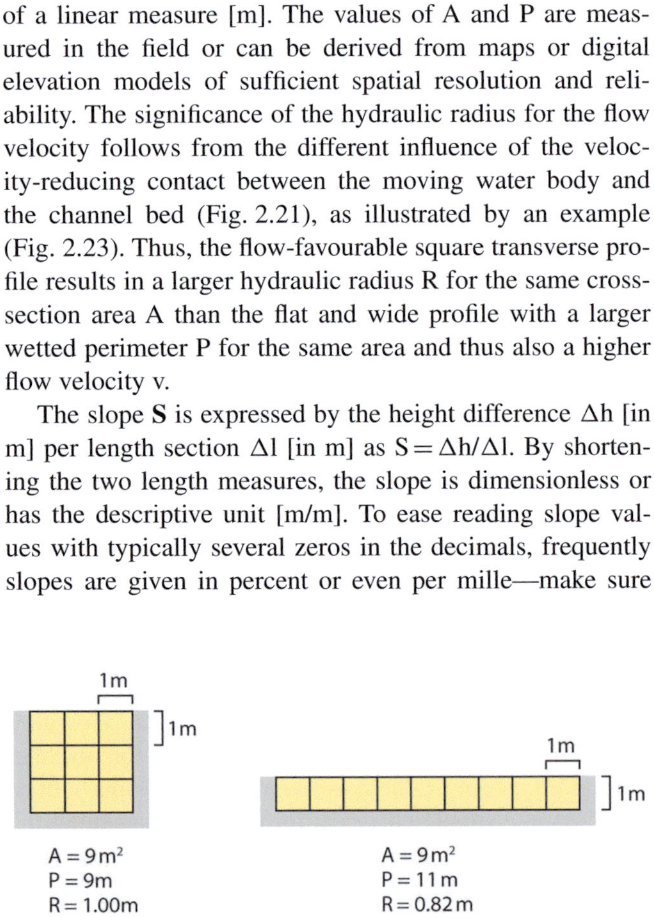

Fig. 2.23 Different hydraulic radius R of two equally sized channel cross-sections A

The determination of the **roughness coefficient n** is more difficult. This value describes the influence of roughness elements such as vegetation, boulders, shallows and pools at the channel bed bottom and banks, which reduce the flow velocity. They all cause an increase in turbulence of the water which has a reducing effect on the mean velocity in the direction of the valley slope. Algebraically, the n-value has the unit [s m$^{-1/3}$], which can be attributed to the empirical character of the formula. The concrete value for n must be estimated, which requires appropriate experience, but is facilitated by aids such as tables (Table 2.4), manuals with instructions on how to proceed (Acrement and Schneider 1989; Chow 1959), or comparative illustrations (Fasken 1963; Barnes 1967; Coon 1998; Hicks and Mason 1998).

In practice, the application of the Manning formula for flood reconstruction has often been criticised (Yen 1992; Ferguson et al. 2017). The flow process is extremely simplified by assuming a one-dimensional steady discharge. With these key hydraulic vocabulary, it is expressed that no eddies in the flowing water to the sides or vertically are taken into account (second and third dimension) and the change of the discharge during the rise and fall of the water level during the passage of a flood wave (unsteady, i.e. changing discharge by time) is also neglected. Other aspects that are physically important for the flow process, such as the viscosity of the water, which is influenced by temperature and sediment content, are also ignored in favour of simple applicability. Thus, if one compares the Manning formula with the processes in reality, it is far from being able to represent them even approximately. On the other hand, one must be aware that for more elaborate models or even for physically based simulations, a multitude of parameters must be quantified and fed in, which can hardly be collected representatively in practice, let alone reconstructed for past floods. Effort and result should be balanced, especially since comparative studies have shown that for elaborate realistic models, a large number of input parameters (water temperature, sediment content, flood hydrograph) can only be assumed and do not offer higher precision in the final result compared to a simple empirical approach (Carling et al. 2003; Alho and Aaltonen 2008). Even though approaches using the Manning formula tend to overestimate the discharge (Quick 1991; Herget et al.

Table 2.4 Roughness coefficients n for natural streams and rivers (modified after Chow 1959)

Type of channel and description	Minimum	Typical	Maximum
1. **Minor streams** (top width at flood stage < 33 m)			
Lowland streams			
Clean, straight, full stage, no rifts or deep pools	0.025	0.030	0.033
Like before, but more stones and weeds	0.030	0.035	0.040
Clean, winding, some pools and shoals	0.033	0.040	0.045
Like before, but some stones and weeds	0.035	0.045	0.050
Like before, lower stages, more ineffective slopes and sections	0.040	0.048	0.055
Clean, winding, pools and shoals, more stones	0.045	0.050	0.060
Sluggish reaches, weedy, deep pools	0.050	0.070	0.080
Very weedy reaches, deep pools, or floodways weith heavy stand of timber and underbush	0.075	0.100	0.150
Mountain streams (no vegetation in channel, banks usually steep, trees and brush along banks submerged at high stages)			
Bottom: gravels, cobbles, and few boulders	0.030	0.040	0.050
Bottom: cobbles with large boulders	0.040	0.050	0.070
2. **Floodplains**			
Pasture, no brush			
Short grass	0.025	0.030	0.035
High grass	0.030	0.035	0.050
Cultivated areas			
No crop	0.020	0.030	0.040
Mature row crops	0.025	0.035	0.045
Mature field crops	0.030	0.040	0.050
Brush			
Scattered brush, heavy weeds	0.035	0.050	0.070
Light brush and trees, in winter	0.035	0.050	0.060
Light brush and trees, in summer	0.040	0.060	0.080
Medium to dense brush, in winter	0.045	0.070	0.110
Medium to dense brush, in summer	0.070	0.100	0.160
Trees			
Dense willows, summer, straight	0.110	0.150	0.200
Cleared land with tree stumps, no sprouts	0.030	0.040	0.050
Like above, but with heavy growth of sprouts	0.050	0.060	0.080
Heavy stand of timber, a few down trees, little undergrowth, flood stage below branches	0.080	0.100	0.120
Like above, but with flood stage reaching branches	0.100	0.120	0.160
3. **Major streams** (top width at flood stage > 33 m)			
The n-value is less than that for minor streams of similar description, because banks offer less effective resistance			
Regular section with no boulders or brush	0.025	…	0.060
Irregular and rough section	0.035	…	0.100

Note that the values characterised as typical in Table 2.4 do not represent simple mean values in the range of the respective description, but form weightings within a range. Furthermore, the values of neighbouring classes overlap. This simplifies the classification of the terrain findings, as interpretations of the descriptions are suggested and not static classifications are prescribed. In cases of doubt, it has proven useful to carry out the further calculations in parallel with the two values of a plausible range for the estimated roughness

2014), sensitivity analyses show that the influence of the uncertainty in the essential factor of the n-value-estimation is not greater than the uncertainty in other indirect discharge estimates if a careful approach is taken (Wohl 1998) and that the evaluation of gauge data is certainly problematic, especially in the case of extreme flood levels (e.g. Potter and Walker 1985). In the meantime, many adaptations of the formula and approaches to quantifying the parameters have been developed (Yen 1992), some of which are mentioned below and some of which are discussed in more detail in subsequent chapters.

The estimation of the roughness coefficient n can be simplified considerably if the cross-section area is divided into homogeneous units of uniform hydraulic roughness such as channel bed, floodplain and floodplain channels (Fig. 2.4). The flow velocities and resulting discharges are calculated individually and subsequently added up to the total discharge. The sub-elements of hydraulic roughness in these units of the cross-section area can also be analysed individually and added to an overall roughness coefficient for each unit (Chow 1959; Herget et al. 2014). This division of roughness into sub-areas is explained in Chap. 3 by the example of the reconstruction of historic flood discharges in Cologne.

Dissatisfaction with the necessary estimation of the roughness coefficient n has led to the much-cited approach of considering the grain size of the sediments at the channel bottom as an essential element of hydraulic roughness (Limerinos 1970; Bathurst 1978). This approach is particularly useful for streams with shallow water depths, because at higher water levels the influence of grain size at the bed and bank is reduced in favour of bed forms such as pools, shoals or bars. In this context, indirect discharge estimates in mountain streams should also be mentioned, where the water flows between and over boulders and thus the cross-section area is difficult to determine. Since under such circumstances the roughness elements can hardly be classified in a representative and transferable way, more extensive empirical values apply in this context (Yen 1992; Marcus et al. 1992).

Even though the Manning formula is not developed for very small channels such as erosion rills or overland flow, there is currently no applicable alternative despite intensive work in erosion research (Smith et al. 2007).

The opposite extreme of the range in the application of flow velocity estimation by the Manning equation is found in the reconstruction of extra-terrestrial flows, for example on Mars. Even if the causes, and even in some cases the flowing medium, differ from the water flows on Earth, they are nevertheless comparable in principle (Burr 2010) and can be estimated by taking into account the different densities and the different gravities by integrating appropriate correction factors (Carr 1996) with approaches derived from the Manning equation.

2.3.2 Interpretation of Transport Competence by Grain Size of Sediments

Even instinctively, when looking at flood deposits, one recognises a relation between the grain size of the sediments and the energy content of the water that transported them: large blocks can only have been transported at high flow velocities, while the deposition of the finest sediments requires almost still water (Fig. 2.24).

The ability to transport coarse particles is called transport competence. It is a function of the drag force of the flowing water, and can be correlated accordingly well with this, the flow velocity, the water depth or the discharge (e.g. Costa 1983; Komar 1989; Williams 1983). The closer and more direct the physical and hydraulic correlation, the better and clearer the correlation. The hydraulic correlations are deliberately not dealt with in detail here (cf. among others Costa 1983; Chanson 1999). However, it is important to understand some details in order to be able to understand the limitations of this approach to reconstruct hydraulic conditions, namely flow velocity. In general, in connection with the transport competence, it is important to distinguish

Fig. 2.24 Channel bed of the Landwasserbach, Grisons, Switzerland. Rounded boulders are evidence of the drag force of the stream during floods, while on the far bank fine-grained sands have been deposited in the current shadow of the bedrock outcrop (flow direction to the right)

whether the particles are transported as bedload rolling and sliding along the bed, or freely suspended as suspension load, because in this case the competence is indicated at different depths (Turowski et al. 2010). The identification of the transport mode can be derived from the local topographic respectively bedding condtions (Fig. 2.25).

However, before transport comes erosion. In order to erode a particle at the bottom of the channel bed, the shear force resulting from the flow velocity must be greater than the inertial forces. The inertial forces consist on the one hand of gravity, resulting from the weight of the particle, and friction between individual particles, which depends on the surface properties. In the case of fine sediment, there is also the cohesion between the particles, which will not be discussed further here as the focus is on coarse sediments indicating the transport competence. The gradient of the flow velocity between the upper and lower edge of the particle generates a lifting force that facilitates the lifting of the particle out of the deposit. Vortices resulting from turbulence on the downstream side of the particle support the lifting force. Except for very steep channels ($S>0.1$), experience shows that the slope of the channel bed as a factor of facilitated erosion is negligible.

What looks like an obvious interaction of forces and should supposedly be easy to calculate, in practice shows problems in the physical solution, without incorporating empirical experiences. The problem lies in the shape of the particles, which can be streamlined if they are the same size, making erosion more difficult, and in the possibility that irregularly shaped particles are strongly interlocked. Therefore, the equations for balancing the forces can only be solved by empirical coefficients (Costa 1983; Komar 1988).

From the extensive literature on the subject, an approach with a focus on coarse-grained deposits is presented here, which is particularly suitable for the simple and direct interpretation of flood deposits. In practice, the largest and most uniform stones and boulders transported by the flood are measured. From these values, a representative mean value is formed, which equalises local flow effects and the influence of the individual block shape. From literature data, where coarse-grained deposits and flow velocities were determined independently of each other, Costa (1983) compiled the data and determined a regression equation between the grain diameter and the mean flow velocity (Fig. 2.26). This flow velocity describes a minimum flow velocity for the displacement of sediments in a river.

For fine-grained sediments up to the gravel fraction, the relation between the dynamics of sediment transport and the mean flow velocity recognised by Hjulström is shown in Fig. 2.26 as a functional equation and graph with the

Fig. 2.25 Rounded and intensively weathered boulder deposit on a terrace level of Katun River, Altai Mountains, Siberia (same location like Figs. 2.35 and 5.10). The bedding conditions visible along the walls of the artificial ditch illustrate a local scour around the boulder which proves, that the boulder was deposited after the formation of the terrace level

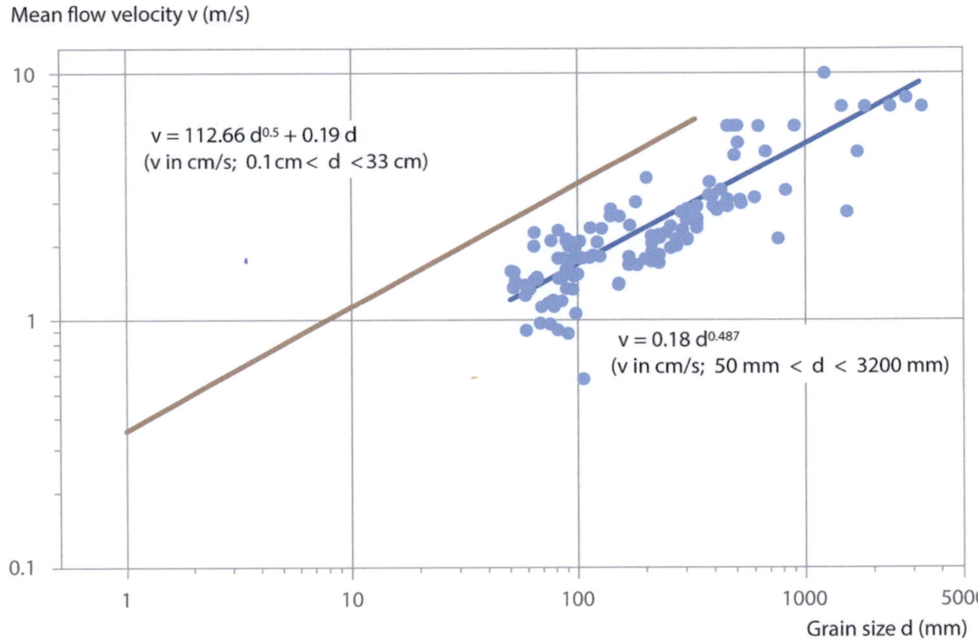

Fig. 2.26 Observations on the relationship between mean grain diameter of eroded sediments and mean flow velocity (Data from Costa 1983 and Zanke 1982)

mean critical flow velocity v_{cm} (Zanke 1982). For non-cohesive sediments up to a mean diameter d<33 cm, this has been shown to be valid and can generally be expressed as

$$v_{cm} = 2.8\left(p'g\,d\right)^{0.5} + 14.7\upsilon/d$$

with v_{cm}—mean velocity to erosion, p′—relative density, g—gravitational constant, d—mean grain diameter and υ—kinematic viscosity. Irrespective of the differentiation according to density and viscosity and by inserting corresponding standard values (p'=1.65; g=981 cm/s^2; υ=0.01306 cm^2/s at 10 °C.), this formula can be generalised in a simplified form as

$$v_{cm} = 112.66\,d^{0.5} + 0.19\,d$$

with v_{cm} in cm/s and 0.1 cm<d<33 cm.

With regard to the position of the curve, it is noticeable that a further extrapolation typically leads to an overestimation of the actual flow velocity required. Such an overestimation is inherent in many physically based equations for the transport of coarse-grained particles (Komar 1989). Therefore, the empirical regression equation of Costa (1983), which was derived from the data set presented as a point cloud and takes into account further empirical values, is introduced for stones and boulders:

$$v = 0.18\,d^{0.487}$$

with v in m/s and 50 mm<d<3200 mm.

For very large grain diameters, the large scatter around the regression line in the double logarithmic scale is also noticeable here; Williams and Costa (1988) estimate the error of the formula at 25–100%. This is probably due to flow anomalies that can occur with high-energy discharges. These are macroturbulence vortex systems that form strong buoyancy forces and thus facilitate erosion and the transport of coarse particles (Matthes 1947; Jackson 1976). Further studies on this are still pending, because the theory-based approaches cannot be applied due to missing data for input variables. Difficulties also arise in practice, since flows in which metre-sized blocks are moved cannot be measured directly. It is also important to ensure that the discharge does not exceed critical threshold values of sediment content, because then the flow laws change due to the grain-supported transport of the particles and coarse grain sizes can be transported more easily and quickly (Costa 1988a, b; Pierson 2005).

Here, limits to the applicability of the competence approach are indicated, which are unfortunately blurred. These are reached in particular when extreme discharge conditions are considered or disproportionate precision is expected. For example, Baker (1973) showed that for extreme water depths of hundreds of metres like they occurred during Pleistocene ice-dammed lake outburst floods, otherwise common laws are no longer valid. Thus, the ratio of mean flow velocity v and erosion-critical flow velocity at the channel bed v_{cs} of $v \approx 1.4\,v_{cs}$ (Zanke 1982), which is typical for water depths of>30 cm, no longer applies. The background to this empirical value is the characteristic vertical distribution of flow velocity (Fig. 2.21), which forms the basis for corresponding statements. Thus, the flow velocity increasing vertically from the bottom and thus also the shear stress at extreme water depths up to the particle upper edge still remain below the threshold value required for erosion. Here, the advantage of considering

Fig. 2.27 Dry outlet of Lake Mashey (Altai Mountains, Siberia). The edge-rounded boulders at the channel bottom originate from the adjacent moraine (left) or rockfall (right), while the runoff in the spillway itself only has the competence to transport fist-sized gravels

the erosion-critical flow velocity at the bottom instead of an average flow velocity over the entire profile becomes apparent.

Problems can also be encountered in doubtless identification of fluvial deposited and redeposited blocks and recording them in a representative manner (e.g. Church et al. 1987; Komar and Carling 1991; Ferguson 2005). It is also common to observe large rounded boulders in river and stream channels that have not been deposited by flowing water. Such blocks may have been deposited by landslides or washed out of adjacent moraine deposits (Fig. 2.27).

Due to weathering processes or previous shaping, they might look like fluvial deposits similar to the surrounding neighbours, thus giving rise to misinterpretations. It is also critical that, especially during floods, the fine-grained base of large boulders erodes, washing away the bottom beneath their feet. These then start rolling and are rounded like fluvial gravel by the suspended load of the water flowing past (Fig. 2.25). Even if this rolling movement is certainly not sufficient for long transport distances, the block nevertheless moved during the flood despite this actually might have insufficient shear stress but produces corresponding traces of fresh rearrangement.

Although there are some problems with the application of the competence approach with regard to hydraulic processes (Komar 1989), under favourable circumstances a remarkable accuracy of flood magnitude estimation can be achieved (Costa 1983). Especially if no or no sufficiently precise flood level indicators are available, the competence approach can score points. This can be illustrated by the example of the Santa Ana River, which flows through the Los Angeles Basin (Clarke 1996). Due to intensive development, the riverbed has been completely reshaped, so that information on historical flood levels cannot practically be used for runoff reconstructions. However, blocks relocated in the upper reaches allow hydraulic reconstructions of previous floods and offer an approach for dimensioning flood retention basins beyond only theoretical water balance calculations. Another example is the flash flood that occurred in the Spanish Pyrenees in 1996, which exceeded all empirical values respectively predicted magnitudes and costed 87 lives. By interpreting the mobilised blocks, it was possible to reconstruct the runoff values and thus adjust the flood forecast for future extreme events (Alcoverro et al. 1999). In a wadi not far from the holiday resort of Sharm El-Sheikh on the Sinai Peninsula, Kehew et al. (2010) were able to verify magnitudes of local flash floods based on the interpretation of block deposits. Their findings contribute to flood protection in an area for which only few observations are available and hardly any alternative more reliable approaches are applicable.

By a closer look, several additional aspects and influences might complicate the interpretation of transport competences as selected examples might illustrate. Greenbaum et al. (2020) concluded for an active mobile sand layer, which supported the transport of boulders in the Negev Desert. Alexander and Cooker (2016) draw attention on the additional effects of the so-called impulsive force of flash floods. The extreme variation of forces during a flash flood event generates significant transient forces in addition to the mean-flow drag force which results in an overestimation of the related transport competence. A similar effect but related to rather slow respectively low energetic flow conditions has the transport of sediments by ice rafts. De Brue et al. (2015) calculated the related ice-rafted transport competence considering the flow velocity of the River Meuse being too slow for the transport of 2 m-boulders due to the low slope without such support. In addition, minimum water depths could be estimated considering the required draught of the loaded rafts.

Compared to the interpretation of the transport competence from the redeposited bed load described above, corresponding hydraulic interpretations of the suspension load are comparatively rare. Their transport results in the widespread alluvial loam of the river valleys, which consists of fine-grained clay, silt and sand aggregates. Under common flow conditions, these grain fractions can be eroded, kept in suspension by the turbulence of the flow and thus transported over long distances. During the transport, the sinking velocity of the particles is lower than the turbulent upwelling and mixing flows. The amount of suspension load is clearly visible by the typical turbulence of the flowing dirty looking water (Fig. 2.28).

When the river leaves the channel bed in the event of a flood and begins to inundate the floodplain, the suspended load it carries is also transported there. As the flood level falls, the effect of roughness, for example due to vegetation, begins to be stronger in the floodplain than in the channel, resulting in increased deposition of the suspended load due to decreasing flow velocity. The deposited maximum grain size is therefore proportional to a minimum flow velocity that must have prevailed at this point during the flood, because otherwise this grain fraction would not have been transported through the corresponding floodplain at all.

From measurements (cf. Zanke 1982), an empirical value of the grain size can be derived, which can be transported in suspension as a function of the mean flow velocity v:

$$d = v^2/(360\,g) \text{ or}$$
$$d = v^2/3561$$

with $g = 9.81\ m/s^2$ where v is given in m/s and d in m. From sandy deposits, which according to their location and depositional structure can only have been transported in suspension, a minimum flow velocity can thus be derived by transforming the above formula. The range of validity of the regression Eq. (0.2 mm $< d <$ 0.85 mm) results in a flow velocity range that is comparatively unspectacular (0.8 m/s $< v <$ 1.9 m/s), since corresponding flow velocities can already be observed in frequent mean floods. However, glacial lake outburst floods in Siberia have transported grain sizes in suspension that exceed the above range of empirical values (cf. Chap. 5).

A wealth of qualitative information on discharge conditions can be derived from the sedimentology of floodplain

Fig. 2.28 Mouth of the River Moselle into River Rhine in Koblenz (Photo: Stefan Harnischmacher). The different suspension load of the rivers is clearly visible, which can be attributed to its reduction by sedimentation in the River Moselle due to low flow velocities upstream caused by the numerous weirs along the river

deposits (e.g. Allen 1965; Anderson et al. 1996; Brown 1997; Bridge 2003). For example, Sundborg and Calles (2001) show how discharges can be estimated from the comparison of grain size spectra of suspension sediments, an approach that was applied to quantify discharges of historic floods of the River Rhine by Toonen (2013) and Toonen et al. (2015 and 2020). Note, that qualitative interpretations of floodplain sediments indicating e.g. wood clearing or other palaeogeographical inferences are not considered here.

2.3.3 Geomorphological Features in the Channel and Floodplain

Assuming sufficient thrust, flowing water can have an erosive effect on the bed. While the previous section focuses which grain fraction of unconsolidated sediments can be eroded and transported, the following section deals with bedforms at the bottom of the channel. Bedforms are formed accumulatively by the redeposition respectively accumulation of unconsolidated sediments or remain as erosional forms.

In general, it is easy to imagine that there are fundamental connections between the discharge conditions, in this case the flow velocity, and the generation of forms on the channel bed and floodplain. Without going into details here, it is obvious that, for example, for scour formation, eddies must exist that require a minimum flow velocity and water depth to develop. Therefore, it can be concluded that hydraulic interpretations of small forms at the channel bed are possible at least qualitatively, and also quantitatively with regard to the exceeding of threshold values. The temporal sequence is problematic, because the bedforms need a certain time to develop to their mature stage, which can take several days for larger forms (Amsler and Schreider 1999; Wilbers 2004). On the other hand, the discharge and thus the boundary conditions such as water depth and flow velocity are unsteady and change continuously during a flood. Thus, erosional forms that have been formed according to the peak discharge can be partially modified when the flood wave subsides due to the sediments that are still carried along (Miwa et al. 1999). Therefore, shape-discharge relationships are not always easily and unambiguously determinable and experiments have often been conducted in flumes to clarify the processes under defined boundary conditions. The technical literature on sedimentology and process research of channel bedforms is extensive, indeed downright confusing. Erosion forms in bedrock have been classified by Richardson and Carling (2005), while Reineck and Singh (1980) or Allen (1984), among others, trace sedimentary structures in the fluvial environment back to their formation processes. In the following, hydraulic interpretation approaches are presented using fluvial ripples and dunes and also local scour at obstacles as illustrative examples.

- **Ripples and dunes**
  Characteristic ripple (Fig. 2.29) and dune shapes (Fig. 2.30) typically form on the bed of flowing waters in sandy-gravelly sediments. Their ridges run transversely to the direction of flow.

Fig. 2.29 Ripple marks in sandy-silty sediments (flow direction to the left)

Fig. 2.30 Fluvial gravel dunes in the Kuray Basin, Altai Mountains, Siberia (flow direction to the left)

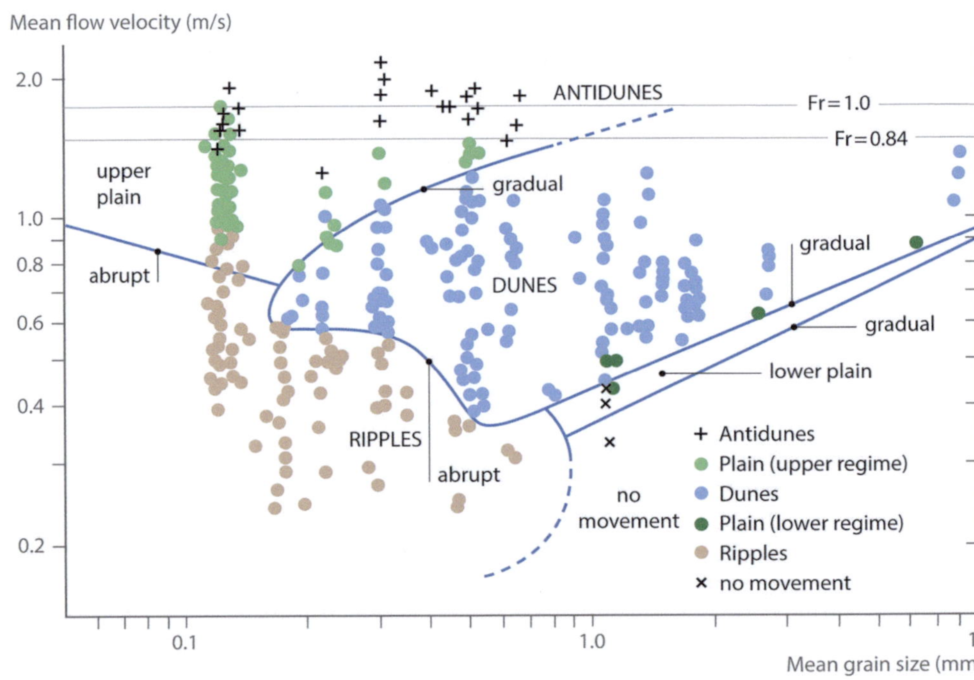

Fig. 2.31 Characteristic stability fields of channel bed forms (depth of flow 0.25–0.40 m; modified after Ashley 1990)

Ripples are rather irregularly shaped and initiated by near-bottom turbulence in the current and therefore do not exceed heights of in the magnitude of centimetres and lengths up to decimetres (Wilbers 2004). Fluvial dunes are wave-shaped with gentle slopes exposed to the current and steep leeward slopes and might reach heights of several metres and wavelengths of several hundred metres (Amsler and Schreider 1999). Classifications are very inconsistently in the literature (Ashley 1990; Wilbers 2004), so that even this basic differentiation between ripples and dunes is not consistently established. Also, the use of the term dune for fluvial forms is frequently less established e.g. in German-speaking countries, but was traditionally limited to aeolian forms. Since ripples corresponding to bed forms can additionally be formed by wave activity and in marine areas with currents in different or several directions—the ripple marks on tidal beaches at low tide illustrate this feature. Their hydraulic interpretation is challenging to generalize (Harms 1969; Allen 1984; Julien and Klaassen 1995). However, the following basic pattern of the different bed-forms and boundary conditions for their formation can be illustrated (Fig. 2.31).

At low flow velocities, ripples already form in fine-grained sediments, whereas in coarser sediments (approx. > 0.7 mm) they cannot form and transport initially takes place over a plane bed. As flow velocity increases, dune formation begins (Fernandez et al. 2006). While the formation and height of ripples is independent of water depth, fluvial dunes can only grow to a maximum value below the water surface: Carling et al. (2000) found values of rarely more than 1 m at around 4 m water depth in the River Rhine near Mainz. Amsler and Schreider (1999) measured more than 5 m dune height at water depths of almost 25 m in the River Parana in Argentina. On the other hand, there is a relation between dune formation and water depth due to the logarithmically increasing flow velocity from the channel bottom to the water surface (Fig. 2.21). If the water is too deep, the flow velocity at the bottom and thus the thrust there is too low to move sediments and form fluvial dunes. The occurrence of fluvial dunes thus indicates maximum water depths at the time of their formation. Such dune scaling relations are intensively investigated and Bradley and Venditti (2017) concluded that frequently a relationship of dune height to depth of flow of 1/6 and dune length to dune height of 5/1 was found while a large range of variability has to be considered. Analogous to aeolian dunes, fluvial dunes form by erosion on the windward side and accumulation on the leeward side, whereby here the slope angles are controlled by sediment properties (Kleinhans 2004). Typically, dunes show a dip of the stratification with the flow direction. Carling (1999) draw attention to special features of gravelly sediments. Ripples and dunes can also complement each other, for example when ripples form on the stoss slope of the dunes, i.e. they are superimposed. With increasing flow velocity, ripples and dunes are again abruptly or gradually receded, whereby again a plain bed is formed, but in a higher energy regime. Anti-dunes are formed at even higher flow velocities or at the transition to supercritical flow with a Froude number > 1 (Carling and Shvidchenko 2002). Here, the channel bed forms run parallel to standing waves on the water surface and can reach heights greater than the water depth. Typically, their internal stratification dips against the direction of flow, caused by accretion at the stoss slope and erosion at the lee slope of the antidunes, distinguishing them from dunes (Carling 1999). As a result, the antidunes migrate against the flow direction of the channel and the slope.

As illustrated in Fig. 2.31, the different bed forms are indicators of the flow velocity during their formation, although there are still open questions about the uniqueness of the relationship (Grant 1997). An important prerequisite for hydraulic interpretation is that the forms had sufficient time under constant discharge conditions to develop to their mature stage. Whether this was the case or the surviving forms were preserved without subsequent alteration by erosion or redeposition can be deduced from the geometric proportions, namely the height/length ratio, for which e.g. Julien and Klaassen (1995) and Carling (1999) cite regressions for different sediments and scales.

While on the one hand there are numerous studies on the formation and dynamics of channel bedforms, this information has so far been little used for the interpretation of palaeoflooods and discharge reconstructions. The reason for this can probably be seen in the fact that only ranges of flow velocities and water depths can be derived from field evidence, which allows only vague estimations. Additionally, the formation of fluvial dunes in nature is rather irregular in terms of geometry (height, wave length, slope properties) and sedimentology (grain sizes, internal structure), which increases the uncertainty. Carling (1996a, b) and Herget (2005) have hydraulically interpreted fluvial gravel dunes of glacial ice dam outbursts in the Siberian Altai Mountains and were able to show that despite impressive size—heights are up to 23 m with wavelengths of over 300 m—they were not formed during peak discharge. The gravel dunes were formed during water depths between 6 and 80 m at flow velocities of 4.8–10 m/s. For antidunes, it could be shown that the flow velocity can be estimated with sufficient accuracy from their wavelength (Froude et al. 2017).

- **Local scour at obstacles**
  Under natural conditions, obstacles such as large boulders, tree trunks and trees protruding from the water are often found on the channel bed of rivers or neighbouring floodplains. The flowing water has to flow around them, which leads to local increases in flow velocity and consequently erosion at the edges: a scour forms. The loose sediments eroded in the scour are redeposited in the flow shadow of the obstacle and form a sediment accumulation there. Together, the scour and the associated sediment ridge form a fluvial obstacle mark (Fig. 2.32).

In early studies, the shapes have been described (e.g. Karcz 1968; Allen 1984), but only to a limited extent explained from the current pattern and related to boundary conditions. Frontal scouring has received particular attention in engineering sciences, as it has repeatedly caused bridges, whose piers have been washed out in the riverbed, to collapse. In these studies, however, only the scouring was of interest, not the entire bedform. Within the framework of experiments in a hydraulic test channel, the flow field around such an obstacle has been systematically investigated (Fig. 2.33) (Euler and Herget 2012; Schlömer et al. 2020).

A downward current forms at the front of the obstacle, which is responsible for the scouring. The water flowing directly past the edge of the obstacle is set into a rotational motion and carries away the eroded sediments. The frontal scouring causes an over-steepening of the scour flanks, which causes sediments to slide in. Subsequently these are also discharged and thus the scour begins to expand further. In the lee of the obstacle, the rotating vortex lanes widen so

Fig. 2.32 Typical fluvial obstacle mark in an experimental channel. In front of the obstacle, a scour has formed on the channel bed. The sediments eroded there have been deposited again in the flow shadow (flow direction to the right, diameter of the obstacle is 3 cm)

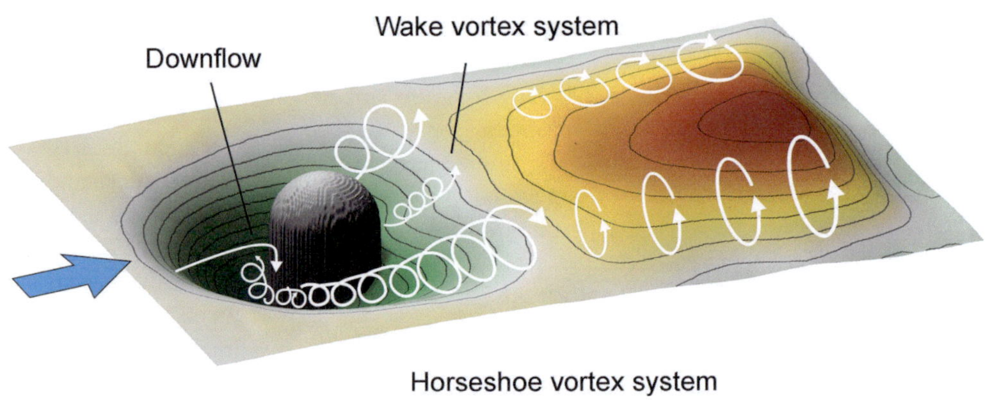

Fig. 2.33 Flow and current pattern in the vicinity of a submerged obstacle

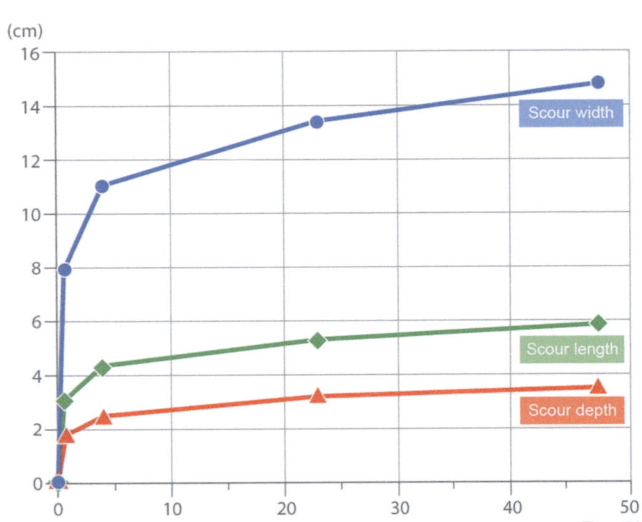

Fig. 2.34 Development of the dimensions of a scour hole in front of a submerged obstacle

that the characteristic horseshoe shape is formed in the plan view. As the drag force decreases, the entrained sediments are deposited and form the sediment ridge. Characteristic for submerged obstacles is the additional lee vortex formation immediately behind the obstacle. The formation of the entire fluvial obstacle mark to a degree of maturity corresponding to the flow conditions takes some time. However, it has been observed that the development is relatively rapid in the early stages and then decreases in intensity. Thus, scour depth and the frontal scour slope develop very rapidly practically up to the degree of maturity, while the scour width controlled by the horseshoe vortex system develops more slowly (Fig. 2.34).

The small-scale investigations in the flume have been verified by comparative recordings in dry riverbeds (Euler et al. 2017). This studies documented, that the scour proportions remain unchanged over several orders of magnitude, so that a transfer of the observed relations seems possible. Compared to other hydraulic approaches—namely the competence approach explained above—this is a particular advantage.

Herget et al. (2013) applied this in the hydraulic interpretation of fluvial obstacle marks generated by Pleistocene ice-dammed lake outburst floods. When comparing the scour proportions in the field with the ratio values in the experimental channel, it can be analysed whether the fluvial obstruction marks were able to develop to the mature stage and are thus meaningful. Figure 2.35 illustrates a fluvial obstacle mark on a boulder that left over from a Pleistocene outburst flood.

The reconstruction of the flow conditions of the fluvial obstacle mark shown in Fig. 2.35 resulted in a mean flow velocity of 2.8–3.5 m/s and an incident flow velocity of 1.8 m/s at the top of the boulder, depending on the unknown water depth (Herget et al. 2013). These values agree well with an approach considering the maximum eroded grain size of 60 mm. According to required flow velocity for the transport competence (Fig. 2.26), this grain size indicates to a mean flow velocity of 2.7 m/s.

After previous attempts to interpret obstacle mark proportions hydraulically were not very successful due to a lack of sufficient basic knowledge (Herget 2005), this approach was also able to verify the consistency of earlier modelling of various events with regard to local water depth and flow velocity. Nevertheless, the approach of hydraulic interpretation of remaining scour structures and fluvial obstacle marks requires further verification. Recent scour structures as well as those left behind by palaeofloods can be found frequently in the field—but they might also show a different pattern. For example, Baker (1973) documents a scour in the wake of a block with a rather gentle inclined front formed by the outburst floods of Glacial Lake Missoula in the Pleistocene. Preliminary investigations in an experimental channel indicate that a strong downward current with resulting scour erosion cannot develop in the case of gently sloping fronts of a submerged obstacle (Euler et al. 2014). Instead, the wake vortex is strengthened and causes erosion behind the obstacle, not accumulation.

2.3.4 Indicators of Energy Transformation

In nature, a variety of phenomena can be observed in which flowing water, due to its inertia or because of energy conservation, undergoes water level changes that are proportional to the flow velocity. Provided that these water level changes are recorded—e.g. by the deposition of flotsam or sedimentary suspension load—they can be subsequently used to reconstruct the flow velocity. In the following, two such phenomena are illustrated and concepts for estimations of derived discharge reconstruction are presented.

- **Superelevation in bends**
  When flowing around bends, the centrifugal force raises the water level on the outside and lowers it on the inside. The resulting difference in height of the water level between the inside and the outside is referred to as superelevation. Influencing variables for the measurement of this superelevation are the radius of the curve and the flow velocity of the water, which results in the equation

Fig. 2.35 Fluvial obstacle mark on a boulder in Katun River Valley (Altai Mountains, Siberia) with schematic representation of the formation. During the Pleistocene ice-dammed lake outburst floods in the Altai Mountains in Siberia (cf. Sect. 5.2), the block was deposited after the bar in the background and the underlying terrace accumulated. In the waning stage of the flood, the deposited boulder acted as obstacle which led to the formation of an obstacle mark. The scour hole, which was up to 2.27 m deep, was subsequently filled in by loess, but the change to the underlying gravels is clearly visible in the trench profile. The sediment ridge in the wake on the right side of the boulder is only faintly distinguishable from the loess filling with a height of up to 0.5 m, but a change in vegetation cover (sparser and lighter in shade) can be seen. The block probably slipped partially into the frontal scour during the flow around it, as its lower edge coincides with that of the scour. The right scheme illustrates shows the area of erosion (-) in front of the block (black) and the accumulation (+), the arrow illustrates the angle of view of the photo

$$\Delta h_w = v^2 W/(g\, R_m)$$

with Δh_w—difference of water surface elevation; v—mean flow velocity; W—width of the channel; g—gravitational constant and R_m—curve radius in the centre line of the channel (French 1994). As the relation is physically based, it can be rearranged and therefore the related flow velocity estimated based on the quantified superelevation—cf. details below. This formula tends to underestimate the water level rise because only the mean flow velocity is taken into account, while in reality significant differences of flow velocities can be observed within a bend (Chow 1959). Thus, the higher flow velocity on the outside leads to secondary flows, from which the well-known phenomenon of accumulation along the point bar and erosion along the cut bank in a river meander can be derived (Naudascher 1992). Furthermore, it is also important to note that the water surface in the cross-section does not form a straight line, but runs parabolically. The relative lowering of the water level in the interior of the curve is greater than its rise at the outer curve. These supposed high-resolution details play a significant role, as the amounts of water level superelevation are in the dimension of centimetres for river widths in several metres. Against the background of these hydraulic details, Apmann (1973) developed a formulation that focuses on the estimation of flow velocity from the measurement of water level superelevation in river meanders:

$$v = (\Delta h_w g/K)^{0.5}$$

where the newly introduced superelevation coefficient K includes the geometric factors resulting from the channel width and the curve radius. It is calculated as

$$K = 1.25 \tanh\,(R_m \Theta/W)\, \ln\,(R_a/R_i)$$

where the hyperbolic tangent is formed from R_m—mean curve radius and Θ—curve opening angle and R_a—radius of the outer curve and R_i—radius of the inner curve (Fig. 2.36).

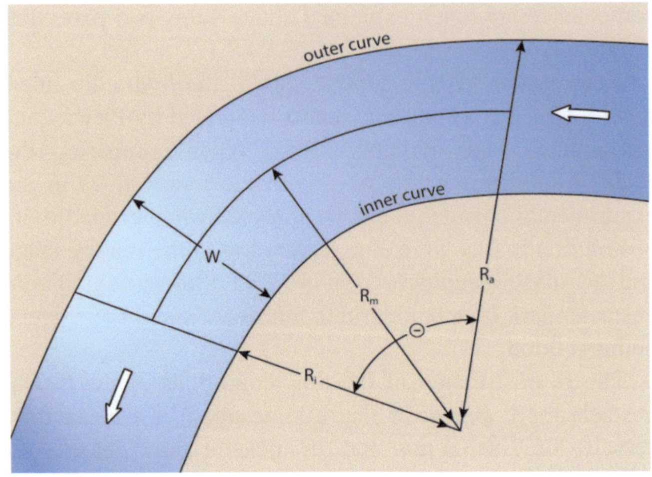

Fig. 2.36 Geometric elements for estimating the water level rise in river curves (abbreviations explained in the text)

As clear and physically justified as the approach is, in practice secondary flows cause problems in its application due to local turbulences and wave formation. Therefore, the calculated individual value should not be overestimated in favour of the formation of a plausible magnitude of flow velocity (Apmann 1973). Obviously, the approach is not applicable for water levels that inundated broad floodplains above and beyond a well-defined river channel. Thorne and Furbish (1995) also had to observe that floodplain vegetation on the banks can permanently disturb the change in water level there due to its roughness and the resulting turbulence.

Even if a variety of complications in the interpretation of water surface elevation in curves should be expected from the theory, this approach is nevertheless frequently applied. For example, Wieczorek et al. (2001) found that the flow velocities of flood discharges during a storm event in coastal Venezuela, which were estimated using the water surface elevation, were more plausible than the values determined using the competence approach. Kershaw et al. (2005) found consistent estimates in comparison with other approaches. This method is frequently used in the investigation of volcanic mudflows, the lahars (e.g. Muñoz-Salinas et al. 2007). An advantage here is that the approach can be applied regardless of the density of the flowing medium and is thus also suitable for mudflows, debris flows and other discharges with an increased sediment concentration (Costa 1984). For these currents, the turbulence of the flowing water is no longer the central parameter for sediment transport, as the sediment content is disproportionately high, so that the "normal" flow laws do not apply and approaches based on them cannot be used. Thresholds for critical sediment content range depend on expectations of accuracy and might start already at > 1–3 volume % (Chanson 1999). Therefore, the approach is particularly suitable for mudflows and debris flows, as high sediment contents are characteristic here but cannot be well-defined quantified.

- **Run-up sediments**
  If the movement of flowing water is abruptly stopped by hitting an obstacle such as a tree trunk or a bridge pier, the kinetic energy must be transferred into potential energy, since it is well known that energy cannot be destroyed. Figure 2.37 illustrates the phenomenon by means of an experiment in a hydraulic test channel.

Despite variable boundary conditions, the phenomenon of water rising in front of a flowing obstacle is related to the bow wave of a ship or the surge of ocean waves in front of a lighthouse or wind generator standing in the water. The amount of this rise in the water level is proportional to the

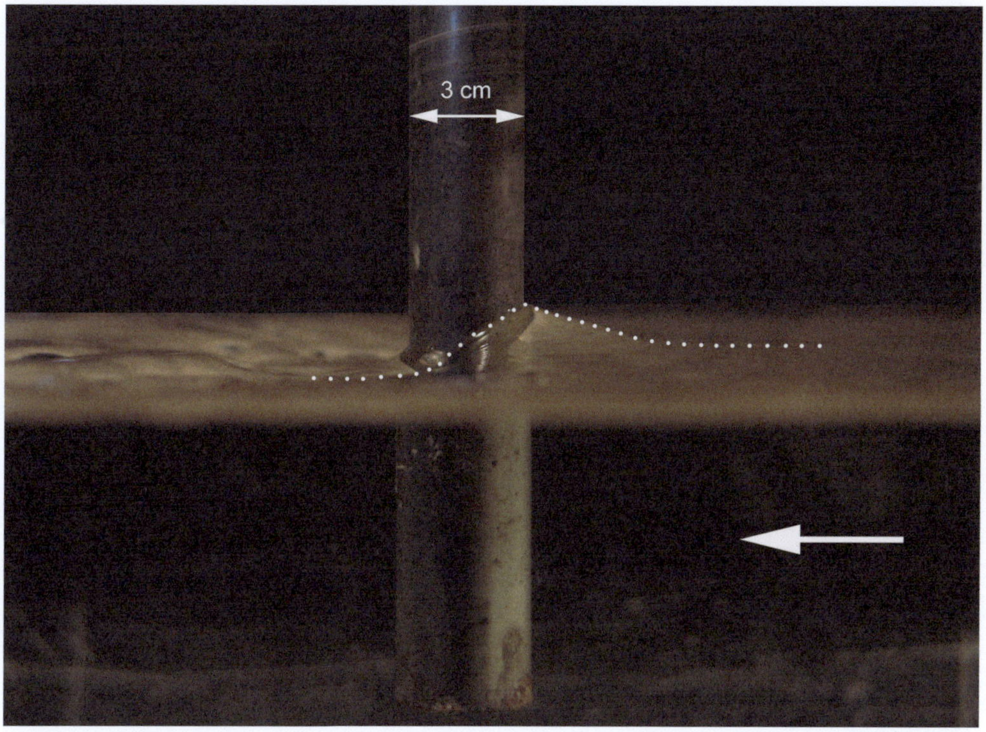

Fig. 2.37 Rise of the water level in front of a cylinder due to the transfer of kinetic into potential energy

flow velocity, which can be easily derived from the general energy equation (Chow 1959; Zanke 1982). Since water is practically non-compressible, this is composed only of the potential height and velocity height, both of which can be specified in units of length, neglecting the pressure energy:

$$H = (y + z) + v_a^2/(2g)$$

with H—constant energy height [in m], y—water depth [in m], z—height above reference datum [in m], v_a—flow velocity [in m/s] and g for the gravitational constant [9.81 m/s^2]. If the entire kinetic energy is now transferred without frictional losses from the incident flow velocity and the energy content H and the reference datum z remain unchanged, the entire energy can only be transferred by a change in the water depth Δy, i.e. a rise in the water level against gravity. Transformed to the incident flow velocity v_a of interest here, the following relationship thus results:

$$v_a = (2g\Delta y)^{0.5}$$

If one also takes into account the non-uniform velocity distribution in a channel by means of an empirical flow velocity coefficient α and uses a typical value of $\alpha = 1.3$ (Chow 1959; Henderson 1966; Hulsing et al. 1966; Li and Hager 1991) and $g = 9.81$ m/s^2 for this purpose, the formula for estimating the mean flow velocity v_m in the channel can be estimated as

$$v_m = (33.2\Delta y)^{0.5}$$

where v_m is given in m/s and the water level rise Δy is given in m. The mean flow velocity is already considered by $v_m = \alpha\, v_a$.

At first glance, it is surprising that the water level rise at an obstacle in a section of the channel cross-section with low incident flow velocity leads to the same flow velocity coefficient α for the mean flow velocity in the overall channel as in areas with high incident flow velocity and correspondingly high water level rise. However, if one considers that an obstacle in a fast-flowing section also leads to stronger turbulence than in a comparatively calm section, the resulting comparable mean flow velocity may already seem plausible. The formation of eddies is clearly visible in the downstream side of the cylinder in Fig. 2.37 by means of the turbulent water surface. In the above-mentioned reference literature, the various factors influencing the flow velocity coefficient are analysed and discussed in more detail.

The phenomenon of water level rise in front of an obstacle is also used as a measuring method of flow velocity (Fonstad et al. 2005). A ruler with water depth markings or a rectangular bar with a favourable flow profile is placed on the channel bed. The difference in water level in front of the rod when the flow-favourable narrow side is exposed and the rise in water level when it turns to face the wider transverse side corresponds to the velocity height Δy to be determined (Fig. 2.38). Multiple measurements and averaging have proved useful to reach the typical inaccuracy of 5%.

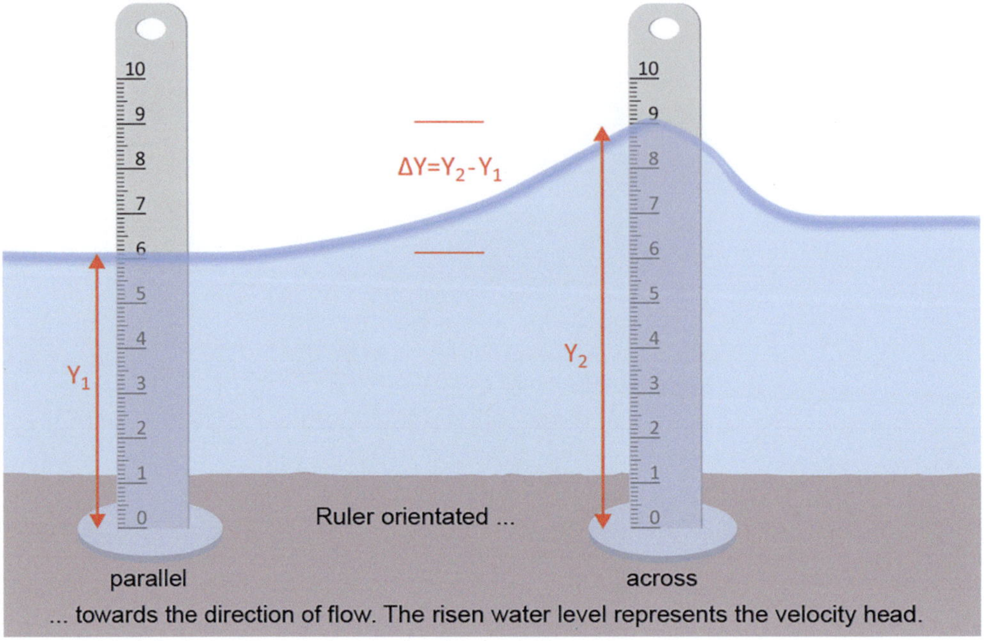

Fig. 2.38 Measurement of velocity height by a ruler parallel and across the direction of flow. The larger area in the across position reduces the flow velocity and results in a risen water level. The amount of the rise represents the velocity head respectively the transfer of kinetic into potential energy

For application in flood reconstruction, the water level upstream of an obstacle and a reference water level must be obtained for comparison. Similar to the water level elevation in curves, the method has therefore been applied for the flow velocity estimation of sediment-rich debris flows, lahars or avalanches (Costa 1984; Pierson 1985), as these leave distinct and differentiated deposits to indicate surface elevations in the field.

In the case of "normal" floods, run-up effects have not yet been used in reconstruction. Here, however, deposits from the suspension freight or flotsam deposits could offer corresponding approaches, if the run-up amount and a reference level can be clearly determined (cf. Fig. 2.19, where this unfortunately less distinct). Alcoverro et al. (1999) mention different flood levels at a building on the flow-exposed and flow-parallel side of 2.0 and 1.2 m, respectively, but did not carry out a more detailed estimation of the local inflow velocity, which might have been almost 3.5 m/s for the difference of 0.8 m. Herget (2005) applied this approach to reconstruct the Pleistocene ice dam outburst floods in the Siberian Altai Mountains and determined discharge values that correspond to those estimated by other methods (cf. Chap. 5 for details). It should be noted that height differences of sediments transported in suspension and sediments deposited in front of rock ridges by run-up were interpreted here, show height differences of > 120 m (Fig. 2.39). Recently, Carling et al. (2020) added theoretical comments on the approach.

2.4 Discharge Modelling

In the previous chapters, the two factors of cross-section area (defined by water level) and mean flow velocity to determine discharge by the continuity equation where discussed. For the estimation of discharge also alternative approaches following very different concepts are developed. While precipitation-runoff-models require input data of sufficient temporal and spatial resolution, black-box models frequently do not consider processes of runoff generation but a purely based on previous experiences and observations. Note, that due to complexity expert models based on e.g. hydraulic modelling or numerical simulations are not considered here.

2.4.1 Runoff Estimation from Rainfall-Runoff Models

The approach of reconstructing historical floods from precipitation-runoff models, analogous to the current real-time flood forecast, is quite obvious. The resulting runoff hydrographs are determined from precipitation data (amount, distribution, intensity), taking into account the general conditions such as e.g. relief, topography and water saturation of the soil. For historical periods, it is possible to reconstruct weather conditions differentiated by atmospheric

Fig. 2.39 Run-up sediments from glacial lake outburst floods in the Chuya Valley, Altai Mountains. The highest run-up sediments are located about 124 m higher than the corresponding deposits of the bar surface in the foreground

circulation, precipitation and temperature at high resolution, for example in Central Europe (e.g. Pauling et al. 2006; Luterbacher et al. 2010; Ljungqvist et al. 2020). In combination with the topographic information from historical sources, precipitation-runoff models can be spatially and temporally differentiated and historical floods can be reconstructed for distinct weather events. The reconstruction of the spatio-temporal distribution of precipitation is problematic, which is especially critical for smaller catchments, as can be observed in the small-scale passage of summer thunderstorms and shower cells.

Bürger et al. (2006) estimated the peak discharge for the extreme flood of October 1824 on the lower parts of the River Neckar, south-western Germany. The approach is based on the determination of analogous current weather conditions for the historical description and reconstructing the resulting flood using the quantified spatial and temporal distribution of precipitation by means of a precipitation-runoff model. As part of the data compilation, historical documents are reviewed with regard to the characteristics of the flood, and current hydrological and meteorological measurement data are collected for the calibration of the model. The time series determined for temperature, air pressure and wind are compiled together with available reconstructions of the large-scale air pressure distribution in historical time to reconstruct historical weather constellations. These are compared with corresponding data based on current records and the associated interpolated precipitation distributions are determined and transferred to historical time accordingly. This provides the input data for the precipitation-runoff model LARSIM and the resulting flood hydrographs can be determined. A further refinement of the approach was carried out by Sudhaus et al. (2008), who modelled an additional flood event with snowmelt components and in parallel also carried out hydraulic peak discharge estimates. The modelled floods correspond well to the available information from historical times, although sufficient meteorological and hydrological measurements must be available for the study area.

Koutroulis and Tsanis (2010) faced the problem of insufficient data for model calibration when modelling flash floods on Crete. However, based on the derivation of the flood hydrograph characteristic for the individual catchment area based on precipitation amount, duration and variability, they were able to reconstruct a flood hydrograph corresponding to the available observations. On closer inspection, this approach is not an universal area-differentiated rainfall-runoff model, as several empirical regressions derived from available data from the study area were interposed. Consequently, the approach is working only within the specific catchment.

In California, significant storms occurred in the winter of 1861/62, which can probably be traced back to a temporarily changed atmospheric circulation pattern (Engstrom 1996). Warm and humid air masses were transported to unusually high latitudes and led to extensive precipitation with resulting floods. In the context of follow-up assessments of natural disasters in California, Porter et al. (2011) used atmospheric circulation models and subsequently a precipitation-runoff simulation to estimate the potential damage from an assumed repetition of corresponding precipitation 150 years ago. Considering that reliable and quantitative records of the historic floods are not available, the simulation remains a pure scenario without any possibility of verification. However, it should be noted, that the focus of the study is on the socio-economic consequences of the catastrophic flood, not its hydrological background.

From the examples above it is obvious, that for the application of precipitation-runoff models for flood reconstruction in historic times, a large amount of spatially and temporally high-resolution data is required, which unfortunately is only available in exceptional cases (Barben 2003). It is also necessary to calibrate the models, especially on extreme events, since non-linear effects play a role, particularly in the case of exceptional floods, and can lead to model failure (Barben et al. 2001). Nevertheless, the concept offers possibilities to conduct plausibility tests of modelling results of extreme precipitation based on historical flood discharges, as some examples might illustrate:

- In the context of the reconstruction of the widespread extreme flood of July 1342 in Central Europe (cf. Chap. 3), Tetzlaff et al. (2002) drew conclusions on the regional precipitation based on the discharge in the River Main. Despite partly unavoidable assumptions and simplifications, conclusion for the effective precipitation of this extreme event could be carried out that are not available from any measured data set for the region. Such data provide a valuable information for worst case scenarios of possible rainfall intensities because of their empirical background—even if exceptional magnitudes are indicated, that might be hard to imagine.
- Reist et al. (2002) pursued a modelling approach similar to that of Bürger et al. (2006) for the reconstruction of a flood that arose from a summer thunderstorm in 1837 in the Emmental, Switzerland. Here, however, the focus was not on the reconstruction of the flood itself, but on the atmospheric boundary conditions. The authors were able to document that previous estimates of maximum area precipitation were clearly too low to produce the runoff of the historical flood even considering different land use.
- Related to this are studies by Pruess et al. (1998) in Colorado, USA. When testing the plausibility of previous flood estimates based on area precipitation and comparing them with flood tracks in the terrain, they were able to determine that there is a prominent upper limit in the mountains for the development of floods from heavy precipitation. Above this, the heavy precipitation cannot contribute to the production of snowmelt water, which is of great importance for the increase in the resulting runoff that would otherwise occur.
- On the other hand, Jarrett and Tomlinson (2000) were also able to show for a catchment in mountainous Colorado that estimates of maximum area precipitation can also lead to significant overestimates of potential flood discharges. Their analysis of all flood traces in the catchment showed that the maximum flood carried only 13% of the assumed potential maximum flood discharge. Thus, the open question of sufficient dimensioning of spillways at dams in the lower catchment could be answered with a previous overestimation of potential maximum discharge. The explanation given for the overestimation is that during the entire Holocene, the expected heavy precipitation cells did not reach the interior of the mountain range, but always rained out at the mountain front beforehand.

2.4.2 Black-Box Models for Runoff Estimation

The counterpart of spatially and temporally high-resolution, process-oriented rainfall-runoff modelling are black box models. Here, no processes are considered, and frequently only causality between the related parameters is considered. In most cases, this relationship is established via regressions, i.e. a purely statistical and non-functional model. What initially sounds quite abstract here is applied in different forms in palaeohydrological modelling, as often the influencing factors cannot be quantified with sufficient precision for process-oriented considerations. Four approaches are presented below. In the context of the overview of methods for runoff estimation, the special approaches for estimating the outburst floods from lakes and reservoirs are already dealt with here in order to present the spectrum of possibilities more clearly; configure Chap. 4 for more details and the background of the phenomenon.

- **Outburst floods—peak discharge and flood duration based on lake volume**
  One cause of flooding can be the failure of dams and the leakage of reservoirs. The problem with many natural reservoirs is that they are often located remotely in high mountain regions that are difficult to access for detailed investigations of the nature and recent conditions of the dam including characteristics of the valley courses as runoff pathways below.

Clague and Mathews (1973) therefore carried out a correlation between the drained lake volume and the peak discharge of the outburst flood on the basis of 10 glacier-dammed lakes. Even though it is conceivable that with a large volume of spilled lake water, the peak discharge of the outburst flood is also large, i.e. a correlation can be assumed, there is obviously no functional relationship between the two parameters. The observed peak discharge obviously also depends on how fast the gap in the dam opens, how steep the valley slopes are below and how far away from the dam the discharge is observed. These factors influencing the runoff velocity and, because of retention effects, the runoff magnitude are not taken into account, which suggests a poor correlation. However, the authors found a correlation of $r^2 = 0.96$, which is surprisingly good and actually suggests a functional relationship for the drained volume and peak discharge (Fig. 2.40).

Beget (1986) suspected that the surprisingly good correlation was due to the extreme value caused by the inclusion of the Pleistocene Lake Missoula outburst flood (cf. Chap. 5). He removed this value from the regression and added other recent ice dam failures influenced by volcanic activity. It is noteworthy here that the resulting regression did not produce any significant change algebraically, apart from the reduced range of validity and the deteriorated quality of the correlation (Fig. 2.40). Such modifications were continued by Costa (1988a, b), who added in particular comparatively small events from the Alps, which further improved the homogeneity of the basic data set, but did not change the resulting function algebraically in the long term either. In addition to these changes based on modified basic data sets, fundamental methodological considerations have also been taken into account in the further development of the correlation. For example, Desloges et al. (1989) show that when logarithm zing the data sets to calculate the regression equation and the subsequent back transformation, a lasting skewness occurs that distorts the result. Even taking this effect into account, no lasting change in the regression equation occurs (Fig. 2.40).

This only becomes the case when the failure mechanism of the ice-dammed lakes is taken into account. Walder and Costa (1996) distinguish between tunnel drainage under or through the ice and other mechanisms such as overflow or mechanical breakage (cf. Chap. 4 on ice-dammed lakes). The regressions resulting from differentiation of the failure mechanisms are clearly further apart and envelop the range of the previous approaches which do not take this process differentiation into account (Fig. 2.40). Strictly speaking, with this differentiation the area of black-box models in the narrower sense is left towards a grey-box model.

Without discussing the arguments and aspects of the above modifications in detail (cf. Herget 2005), one can assume that the drainage mechanisms significantly influence the peak discharge. On the other hand, it is noticeable that including the regression for tunnel drainage by Walder and Costa (1996), the early approaches all led to exponents in the range of 0.64–0.69 (hence ~ 2/3), which is explained by tunnel drainage as most prominent mechanism (Ng and Björnsson 2003) (cf. Chap. 4.2 on lake outburst floods mechanisms for details).

A first approximation for the estimation of the duration of a flood can be carried out quite easily for lake outbursts with a known peak discharge. The restriction to lake outbursts is based on the fact that the flood volume is also known. When a lake breaks out, logically no more water can flow out than the lake contains in total—a trivial but momentous prerequisite. The peak discharge should have been estimated beforehand using one of the methods listed above.

The simple but rather unrealistic approach is to divide the lake volume by the peak discharge in order to determine the minimum duration of the flood. After transformation, this results from the logical consideration that the discharge can only last until the lake volume is exhausted: discharge

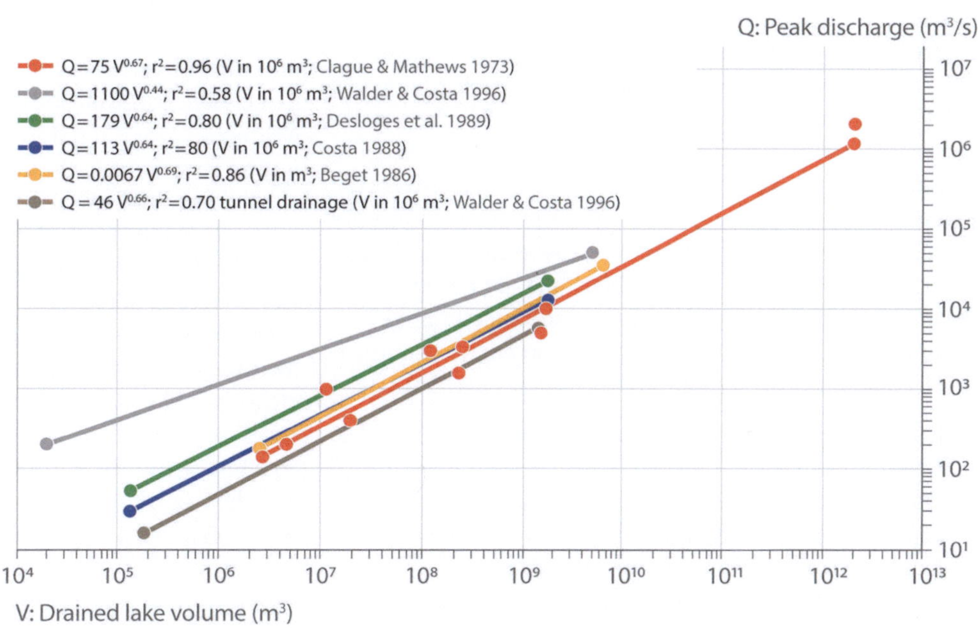

Fig. 2.40 Regressions of the drained volume of ice-dammed lakes and the resulting peak discharge

[m^3/s] × duration [s] = volume [m^3]. Since no specific discharge other than the peak discharge is known, this is used and thus an absolute minimum duration is determined. This sharp characterisation as an absolute minimum value can be made because no hydrograph with a "time-consuming" gradual rise and fall is taken into account. Rather, it is assumed that the drainage instantaneously reached the maximum value when the lake opens and maintains this value without subsiding until it is completely emptied. Such a steady-state peak discharge is of course unrealistic, but the duration of the flood event can also not be less than this value.

In contrast, the construction of an unsteady hydrograph is more realistic. This approach is also based on the value for the peak discharge and the volume of the lake, which is now determined by the integral of the hydrograph curve. Thus, for example, an abrupt rise and a gradual fall can be graphically determined by inserting values on the hydrograph, which are, however, limited in discharge quantity and time by the resulting integral, which must always correspond to the lake volume (Fig. 2.41).

The construction of such a hydrograph is facilitated by various graphics or statistics software with a built-in integral function. However, the control over the integral under the hydrograph leaves less room for manoeuvre than one would expect at first glance, so that there are narrow limits to the design of the hydrograph. Of course, one should not expect any information about the actual dynamics of the flood in detail from such a graphical approach. However, a statement about orders of magnitude of the total duration or exceedance duration of certain discharge threshold values (e.g. discharge at which places are inundated) are possible. In general, however, an underestimation of the total duration is to be expected with this approach, as no retention effects of the runoff can be taken into account. This approach was applied in the reconstruction of Pleistocene glacial lake outburst floods in the Siberian Altai Mountains (cf. Chap. 5). With a lake volume of about 603 km^3 and a peak discharge of about 10,000,000 m^3/s, the duration could be estimated at about 2.5 days (Herget 2005). Given the magnitude of the event, there is no scope for significantly extending the duration of the outflow, as residual discharges form a negligible order of magnitude and are not visible in the graph of the hydrograph. Advanced modelling considering two-dimensional unsteady flow conditions confirmed this estimation in the meantime (Bohorquez et al. 2015).

- **Envelope curves for estimating the maximum possible flood**

  One of the methods for flood estimation that can be applied when neither specific runoff respectively water level nor precipitation data are available is the approach based on the evaluation of empirical envelope curves.

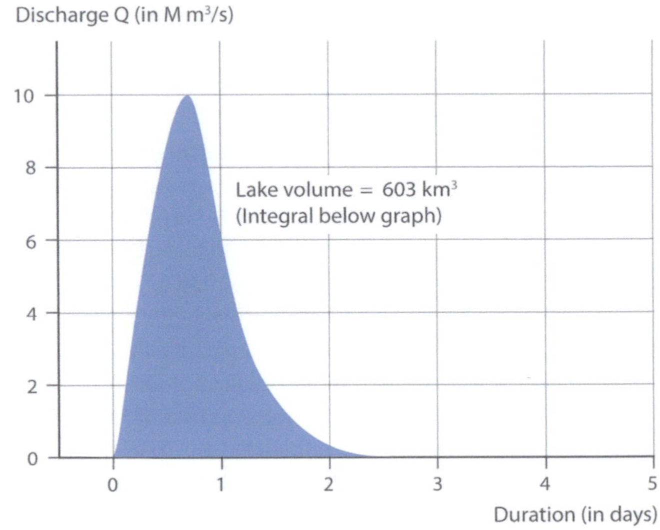

Fig. 2.41 Estimation of the flood duration of a lake outburst flood for known peak discharge and lake volume

Since these critical boundary conditions are also present in the reconstruction of historical and prehistoric floods (e.g. the event is known by its damage but no hydrological data are handed down), this method should be considered here.

Unlike the approaches above, no event-related reconstruction is carried out here, but a maximum discharge for the size of the catchment area is determined. For this purpose, the observed discharges of the catchment areas are converted to discharge peaks per square kilometre of the area in order to achieve a better comparability of the values. Figure 2.42 presents the corresponding discharge values for gauging stations in Germany.

The figure illustrates 986 gauging stations with their largest floods discharges expressed as peak discharge per unit area observed before 1993–95 according to the catchment area classification of the German Hydrological Yearbook (www.dgj.de). Based on similarity of hydrological conditions, these catchments have been roughly subdivided into low mountain ranges with alpine components (Danube, Upper/High Rhine), low mountain ranges with lowland components (Upper Elbe, Main, Middle and Lower Rhine, Weser-/Ems) and pure lowland areas (Havel, Lower Elbe, coastal areas). Due to the location of the catchments and their size, the categorisation is unfortunately not clear-cut. The data set comprising all observed floods is enveloped at the top with a curve, from whose functional equation (here: $Hq = 17{,}000\ A{-}^{0.4}$) the maximum discharge can then be derived as a function of the catchment size, based on previous observations.

The main advantage of this approach is the quick and simple initial estimation of possible flood magnitude in

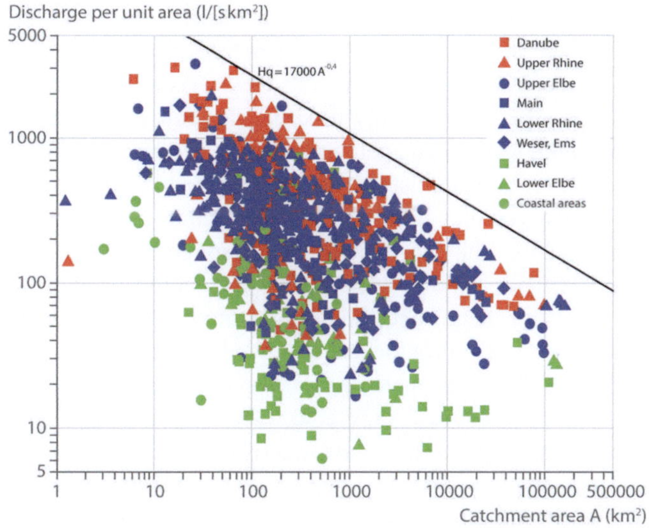

Fig. 2.42 Peak discharges per unit area with envelope curve for different catchments in Germany (data from Hinsberger 2006)

a catchment of a specific size. The estimation via specific discharge is useful because smaller catchments can be completely covered by heavy precipitation cells, resulting in extreme specific discharges that are not possible in larger areas without reaching their extreme peak discharges by summing up the inflows. This is characteristic for the climate in Central Europe, whereas in other climatic zones much larger specific runoff peaks can be observed (e.g. Crippen and Bue 1977; Crippen 1982; Costa 1987; Eide et al. 2018; Herschy 2003; Milliman and Farnsworth 2011; Global Runoff Data Centre www.grdc.bafg.de) and the influencing factors must be differentiated (Hayden 1988; McMahon et al. 1992; Saharia et al. 2017). Disproportionately reduced specific discharges can also be observed, for example in arid regions where a large part of the runoff evaporates. Even within Central Europe, a more far-reaching regionalisation makes sense when drawing up envelope curves (Wundt 1965), as otherwise high runoff yields, such as can be observed at the margins of the Alps, would also be transferred to the lowland waters of northern Germany, where they would lead to a significant overestimation of the possible flood discharges. Of course, such kind of necessary regionalisation is also required in general (e.g. Vogel et al. 2001; Bertola et al. 2023). Considerations of reducing the envelope curve to a lower level of 90% or 50% of the extreme values of the previously observed maximum specific discharge are carried out for not to distort the estimation of expected floods by local and rare extreme events. Wolman and Costa (1984) document in this context that the envelope curves grow with increasing observation duration, as more extreme events are included in the observation series. Note, that in addition to this, extreme events from historic times are usually quantified to estimate potential maximum flood magnitudes—contradictions in the expectations of the flood investigations become obvious.

As simple as the establishment of envelope curves may look, it requires a high level of expertise and knowledge when selecting the data sets to be considered. This can be illustrated by the examples of karstified catchments, where the underground catchment area might exceed the surface catchment size many times. Also discharges from dam breaches as a different cause of a flood event might disturb the consistency of datasets for comparison.

- **Channel geometry as discharge indicator**
  Obviously, there are correlations between the size of a river channel and the discharge: a trickle does not carry discharges in the dimension of several cubic metres per second and a narrow meandering meadow stream does not reach high flow velocities. Assuming appropriate measurement and modelling concepts, these relationships can be calculated today on a physically based basis via several intermediate steps and the associated partial flows can be modelled three-dimensionally (cf. Ikeda and Parker 1989). However, since the effort required for this is quite high, regressions for estimating discharge parameters from channel bed geometry are still used, even though their heyday came to an end in the 1980s, when computer technology opened up a wide range of more advanced calculation possibilities (Thornes 2003).

As the annotated overview by Williams (1984a) illustrates quite well and clearly, a plethora of regressions have been developed, some of which have regionally limited validity, others claiming to be universally valid (Park 1977a, b; Ethridge and Schumm 1978).

Initially, formulas were developed to check the geometric proportions of a river course. For example, width-to-depth ratios of the channel or meander proportions—such as radius to wavelength—are checked to determine whether the river course is in a state of morphological equilibrium or whether it is in a phase of adaptation and transformation due to tectonic uplift or anthropogenic influence (Williams 1986; Wohl and Wilcox 2005). From this, for example, forecasts for valley floor development can be derived, such as the progressive widening of river meanders (Nicoll and Hickin 2010), which can be important for settlements along river courses, or guiding principles for the near-natural design of stream courses can be developed. In addition, there are numerous studies in which channel bed geometry is related to discharge characteristics or flow velocities (e.g. Williams 1984a; Wharton 1995a/b). These regressions are interesting for the reconstruction of discharges, which is why they are examined in more detail here.

As with the application of envelope curves, no event-related flood reconstruction is possible when interpreting the geometry of a channel bed. Rather, this approach looks at the channel bed-forming discharge. It is widely accepted that this is the bankfull discharge Q_b. Even though there are several definitions and approaches to determining this in detail (e.g. Williams 1978; Harnischmacher 2002; Slattery 2015), the bankfull discharge is reached at a water level that lies just below the inundation of the wide floodplain of lowland rivers. Against the background of the fine-grained bank areas of rivers and streams, the exact measure is more difficult to define in practice than one would expect, which explains the various definitions of, for example, the smallest width-to-depth ratios of the channel bed cross-section. For practical application, it is therefore advisable to use, for example, the mean value from three representative profiles for calculations. The maximum discharge that completely fills the river channel is reached at the full water level, which can thus have a decisive bed-forming effect and shape the dimensions of the channel. This channel-bed-filling discharge is an average value that is repeatedly exceeded, but also undershot, which is why Kochel and Baker (1988) assign this approach to the determination of the parameters to average hydrology and not to extreme value estimation. The frequency of occurrence of bankfull discharge has been much and controversially discussed. Many times, the literature gives recurrence intervals of 1.58 or 2.33 years, which has not yet been sufficiently regionalised (Crowder and Knapp 2005), because it is not plausible that all runoff regimes on earth—especially in areas with only episodic runoff—should follow the same pattern. In the context of a literature review on this question, Nadolny (1994) came to the working hypothesis that the return interval of the bankfull discharge is usually one to two years, deviations from this indicate special cases and streams should be considered in a more differentiated way in this question. On the other hand, regressions have also been developed without the functional consideration of bankfull discharge, whereby the regional limitation of validity is explicitly pointed out in this context (Wharton 1995a, b).

As the few examples in Table 2.5 illustrate, the approach has found application on various occasions, even if it is received with scepticism because of the unclear functional relationships (e.g. Williams 1984b). Anyhow, the advantages, especially the efficiency of the estimation, require further attention. For example, Bjerklie (2007) illustrates that numerous channel geometric parameter can also be derived from remote sensing data. In addition to current discharge measurements via satellite images (e.g. King et al. 2018; Oubanas et al. 2018; Pavelsky 2014), basic discharge estimates can thus be made even for inaccessible areas and without the installation and long-term operation of gauges.

For the catchment area of the River Tisza in Hungary, which has been intensively modified by cultivation, settlements and water management, Gábris (2001) and Timár and Gábris (2008) illustrate how estimations on the natural discharge dynamics in historical times can be reconstructed from data on the wavelength of meanders of floodplain channels and oxbow lakes. This estimation cannot be derived from gauge data due to the fundamental transformation and thus provide an independent estimate of the adequacy of the dimensioning of flood protection. Also for Tisza River, Cserkész-Nagy et al. (2012) derived the mean discharge around 45,000 years ago from the texture of point bar deposits, thus closing a gap in the knowledge about the differentiated discharge dynamics in the unglaciated regions of Europe in Pleistocene times.

However, channel beds have also been preserved from prehistoric times, which contain information about the temporal development of runoff dynamics. For example, the climate change at the end of the last glaciation resulted in a change of the channel pattern of the rivers from a braided (Fig. 2.43) to a meandering river bed (Fig. 2.44).

While the braided channel bed pattern is characteristic of a pronounced annual runoff cycle with significant floods and comparatively low mean discharge, a meandering river course represents a more moderate discharge regime (Vandenberghe 2003; Kaiser et al. 2012). On closer examination, the transition from the glacial braided to a Holocene meandering channel pattern shows spatial differentiation and fluctuations corresponding to climatic setbacks, but also reveals the phenomenon of large meanders, which is of particular interest here (Vandenberghe and Sidorchuk 2020). This is already a warm-period flow regime, but the meander arcs were 5–15 times larger than those observed in recent times. Examples can be found throughout Europe.

Quantitative palaeohydrological runoff reconstructions based on meander geometry were carried out e.g. by Rotnicki (1991) and Starkel (1996) for the Polish examples and Sidorchuk et al. (2001) and Borisova et al. (2006) for the Russian ones. For the Polish examples regression equations developed with data from other regions are used in the hydrological interpretation of channel bed dimensions. For the Russian catchments regional characteristics are used by developing specific regressions according to a so-called palaeohydological analogy approach (Sidorchuk and Borisova 2000). For the discharge characteristics sought at the time of the activity of the large palaeo meanders, present-day river systems with comparable climatic conditions as in the past are sought and the corresponding parameters (discharge, flood frequency, discharge volume, etc.) are transferred or estimated by regressions.

Table 2.5 Selected examples for the interpretation of discharges from channel geometry data

Region	Regression	Correlation r^2 Standard deviation Sd (%)	Validity	Comments	References
Sweden	$MQ = 0.06\ B^{1.66}$	$r^2 = 0.86$; Sd = 39	$1.8 < B < 67$	n = 19; $B \approx B_b$	Williams (1984b)
Sweden	$MQ = 0.025\ R_m^{1.58}$	$r^2 = 0.66$; Sd = 70	$8 < R_m < 169$	n = 19; for angle of curvature > 60° (cf. Figure 2–36)	Williams (1984b)
Hungary, Tisza	$MQ = 0.0009\ (L_m/2)^{1.8}$	r^2 = -; 18 < Sd < 57	$1066 < L_m < 4080$	n = 200; floodplain channels and abandoned channels of Tisza River	Timár and Gábris (2008)
USA, Missouri	$MQ = 0.027\ B_b^{1.71}$	$r^2 = 0.86$; Sd = 79	$0.8 < B_b < 430$	n = 252, additional consideration of grain size at channel bed	Osterkamp and Hedman (1982)
Sweden	$MHQ = 1.0\ B^{1.3}$	$r^2 = 0.77$; Sd = 43	$1.8 < B < 67$	n = 19; $B \approx B_b$	Williams (1984b)
Sweden	$MHQ = 0.28\ R_{cm}^{1.38}$	$r^2 = 0.74$; Sd = 48	$8 < R_{cm} < 169$	n = 19; for angle of curvature > 60° (cf. Figure 2–36)	Williams (1984b)
United Kingdom	$MHQ = 0.20\ B_b^{1.97}$	$r^2 = 0.78$; Sd = 27	$3.7 < B_b < 54.8$	n = 75	Wharton (1992)
United Kingdom	$MHQ = 1.16\ A_b^{1.31}$	$r^2 = 0.73$; Sd = 30	$1.7 < A_b < 62.5$	n = 75	Wharton (1992)
United Kingdom	$MHQ = 0.30\ B_b^{2.31}(B_b/T_b)^{-0.48}$	$r^2 = 0.80$; Sd = 26	$3.7 < B_b < 54.8$ $0.5 < T_b < 1.1$	n = 75	Wharton (1992)
United Kingdom	$MHQ = 0.29\ A_b^{1.15}(B_b/T_b)^{0.67}$	$r^2 = 0.80$; Sd = 26	$1.7 < A_b < 62.5$ $3.7 < B_b < 54.8$ $0.5 < T_b < 1.1$	n = 75	Wharton (1992)
USA	$Q_b = 4.0\ A_b^{1.21}\ S^{0.28}$	$r^2 = 0.96$; Sd = 42	$0.7 < A_b < 8510$ $0.00041 < S < 0.081$	n = 233	Williams (1978)
United Kingdom	$HQ_{1.5} = 0.13\ B_b^{2.02}$	$r^2 = 0.78$; Sd = 27	$3.7 < B_b < 54.8$	n = 75	Wharton (1992)
United Kingdom	$HQ_{1.5} = 0.87\ A_b^{1.34}$	$r^2 = 0.72$; Sd = 31	$1.7 < A_b < 62.5$	n = 75	Wharton (1992)
United Kingdom	$HQ_{1.5} = 0.19\ B_b^{2.32}(B_b/T_b)-^{0.43}$	$r^2 = 0.79$; Sd = 26	$3.7 < B_b < 54.8$ $0.5 < T_b < 1.1$	n = 75	Wharton (1992)
United Kingdom	$HQ_{1.5} = 0.19\ A_b^{1.16}(B_b/T_b)^{0.73}$	$r^2 = 0.80$; Sd = 26	$1.7 < A_b < 62.5$ $3.7 < B_b < 54.8$ $0.5 < T_b < 1.1$	n = 75	Wharton (1992)
United Kingdom	$HQ_5 = 0.28\ B_b^{1.93}$	$r^2 = 0.76$; Sd = 27	$3.7 < B_b < 54.8$	n = 75	Wharton (1992)
United Kingdom	$HQ_5 = 1.56\ A_b^{1.30}$	$r^2 = 0.72$; Sd = 29	$1.7 < A_b < 62.5$	n = 75	Wharton (1992)
United Kingdom	$HQ_5 = 0.43\ B_b^{2.29}(B_b/T_b)-^{0.52}$	$r^2 = 0.79$; Sd = 26	$3.7 < B_b < 54.8$ $0.5 < T_b < 1.1$	n = 75	Wharton (1992)
United Kingdom	$HQ_5 = 0.43\ A_b^{1.15}(B_b/T_b)^{0.63}$	$r^2 = 0.79$; Sd = 26	$1.7 < A_b < 62.5$ $3.7 < B_b < 54.8$ $0.5 < T_b < 1.1$	n = 75	Wharton (1992)

A_b: Cross-section area at bankfull discharge (m^2)
B: Width of channel (m)
B_b: Width of channel at bankfull discharge (m)
$HQ_{1.5}$: Flood discharge of a return interval of 1.5 years (m^3/s); probably represents bankfull discharge Q_b
HQ_5: Flood discharge of a return interval of 5 years (m^3/s)
L_m: Wavelength river meander (m)
MHQ: Mean annual flood discharge (m^3/s)
MQ: Mean discharge (m^3/s)
Q_b: Bankfull discharge (m^3/s)
R_m: Mean radius of several river meanders (m)
S: Slope of the river channel, assumed to be similar with slope of energy line (m/m)
T_b: Mean channel depth at bankfull discharge (m), calculated as $T_b = A_b/B_b$
Considering possible accuracy of the estimations, a balance between clarity of the input data—e.g. radius of a complex river meander, channel width, slope of abandoned river meanders—and the functional respectively physical context of the parameters should be taken into account (Dury 1985). Considering more than one correlation and achieving a range of values based on different input data is favourable. Note, that frequently input data related to bankfull discharge are considered while bankfull discharge itself is not estimated by the correlation. Obviously, a functional respectively physical context is considered but the correlation itself is a statistical regression

Fig. 2.43 Braided channel pattern on a glacial outwash plain, Tujuksu-Glacier, Kazakhstan

Fig. 2.44 Meandering stream course in the Eifel Mountains, Germany

It is remarkable that no palaeohydrological reconstructions have yet been carried out for the valley meander courses of rivers, which are widespread in Central Europe (Dury 1985). These valley meanders indicate the course of the originally freely meandering river before a drastic deep erosion impulse occurred due to tectonic uplift of the mountains. In the process, the geometry, namely the wavelengths and radii of the meanders, was retained and transferred to the entire valley course—i.e. not only the course of the river on the valley floor. Accordingly, this phenomenon is also called forced meandering. These valley meanders are typical for numerous rivers in Central Europe, although usually not as dramatically pronounced as on the San Juan River, a tributary of the Colorado River in the south-eastern part of the US state of Utah (Fig. 2.45).

One possible explanation for the fact that no corresponding palaeohydrological meander interpretations have yet been carried out can be seen in the lack of calibration data of the geometry parameters for the discharge at the time before the incision in the late Tertiary (i.e. more than 2.6 million years ago). Obviously, no regionalised data can be available for this, so that an uncertainty and imprecision factor would impact the estimation. On the other hand, there are several palaeohydrological interpretations of deposits and channel parameters from much older times carried out. For example, Padgett and Ehrlich (1976)

Fig. 2.45 Valley meanders of the San Juan River at Goosenecks State Park, southeastern Utah, USA

document how palaeodischarges can be reconstructed from deposits of a meandering channel from the Carboniferous period (about 300 million years ago) and discuss the reliability of the estimates self-critically.

2.5 Field Work for Flood Assessment

The abundance of the possible methodological approaches raises the question of how to practically identify past floods in the field and record the relevant parameters. Existing method-orientated reviews on the investigation of palaeofloods (e.g. Jopling 1966; Benson and Dalrymple 1967; Benito and Thorndycraft 2004) rely on the existing experience of the researcher with regard to methods and techniques of fieldwork. They frequently focus on warnings and discussions of potential problems.

Pruess et al. (1998) have compiled relevant flood indicators from their work in catchments in the Rocky Mountains, USA. The aim is to identify respectively assess the significance of floods rather than the quantification of specific flood events. They differentiate the flood indicators according to the areas of sediments, channel bed, vegetation and age of fluvial deposits by soil formation and litterfall cover (Table 2.6).

When using tables like Table 2.6, one should critically check whether the recording of findings does not become too static and whether special cases are taken into account appropriately and are not ignored or overrated due to a lack of relevant input fields. On the other hand, the evaluation of the scores takes into account that inappropriate rubrics can remain unprocessed and do not lead to distortions in the comparison. It is obvious that an adaptation for individual application, for example for lowland regions, would be appropriate.

Williams and Costa (1988) give practical advice for the recording of recent flood traces. In addition to a list of equipment and instructions for the representative recording of profiles and the corresponding sampling of flood deposits, tips are also given for careful photo documentation in order to subsequently have evaluable photographs available. It is clear from the list of investigations that an emphasis on the hydraulic interpretation of floods is given (modified after Williams and Costa 1988)

1. Because of the different hydraulic properties, it is important to check or exclude that there is no debris flow investigated.
2. Survey of water level indicators to determine the cross-section area.
3. Survey of the water level width to determine the cross-section area.
4. Survey the slope of the flood level using the remaining flood level indicators and the recent water level in channel without steps or backwater.
5. Investigation of the roughness coefficient n according to the Manning formula.
6. Representative recording respectively sampling of grain size, distribution and volume of flood deposits.
7. Recording including survey of relevant sedimentological structures in the channel bed and the floodplain.
8. Record the lithology and shape of the coarser deposits to estimate the transport distance by comparison with the areas of origin/distribution of corresponding bedrock from geological maps.
9. Investigation of the condition and composition of vegetation to estimate flood recurrence intervals.
10. Recording of further parameters for estimating the mean or local flow velocity and discharge.

The two authors' guidance is formed by their experience of the need to apply indirect discharge estimates in the USA, where the gauging network is comparatively sparse and a historical record of flood magnitudes is mostly lacking. Accordingly, they caution recording traces early after a flood event, as numerous details can be irretrievably lost relatively quickly during clean-up and due to vegetation cover. On the other hand, it is obvious that the tools for recording historical and prehistoric floods hardly differ fundamentally and only the type and extent of the surviving flood traces are reduced for evaluation and interpretation.

To compile a flood chronology of a specific river for historic times in settled areas an evaluation of anthropogenic sources (cf. Table 2.1) has priority over field work beyond searching for historic flood marks on buildings. Typically, the search for historical documents in archives is only a second step after a screening of the local respectively regional history compiled in previous publications by professional or local historians. While this is a rather easy task for prominent cities and locations it becomes challenging in the rural countryside. Prominent locations have to be identified first as the potential for any published history and historic flood marks on buildings is concentrated here—monasteries, castles and seats of local governments illustrate such locations. Note, that prominent locations can also be remote water

Table 2.6 Checklist for flood frequency and intensity in mountainous areas (modified and simplified according to Pruess et al. 1998)

Location, catchment, topography:

Location, catchment: ________________

Size catchment: ________________

Land use within catchment (recent / historical):

____

____

____

Impact on channel bed (intensity, activity):

____

____

____

Elevation: ________________

Slope: ________________

Channel width and depth: ________________

Channel pattern: ________________

Comments:

____

____

____

____

____

Rank the items in the boxes below for each site investigated. Cumulative site scores can then be compared to assess relative flood histories and magnitudes.

Sediments:

Well sorted, mostly fine grained	1	2	3	4	5	Poorly sorted, mostly coarse grained
Similar grain size at tributary junctions	1	2	3	4	5	Increase in grain size at tributary junctions
No traces of soil erosion on hillslopes	1	2	3	4	5	Soil erosion on hillslopes
Limited bars, finegrained	1	2	3	4	5	Flood bars, coarse grained
Rare slackwater deposits	1	2	3	4	5	Frequent slackwater deposits
Round clasts in stable position	1	2	3	4	5	Angular clasts recently moved
Cobble to small boulders moved	1	2	3	4	5	Large boulders moved

Total points for sediments: ____________

Channel form:

Table 2.6 (continued)

Little channel scour	1	2	3	4	5	Channel scour
Untruncated colluvium or banks	1	2	3	4	5	Truncated colluvium or banks
Unincised channel	1	2	3	4	5	Incised channel
Sparse scour and deposition	1	2	3	4	5	Alternating scour and deposition
Controlled by large woody debris	1	2	3	4	5	Channel clear
Unscoured tributaries	1	2	3	4	5	Scour in tributaries

Total points for channel form: ____________

Vegetation:

Lichen cover on exposed rocks	1	2	3	4	5	Lichen cover stripped / missing
Algae cover on submerged rocks	1	2	3	4	5	Algae removed / missing
Old vegetation on banks	1	2	3	4	5	Young vegetation on banks
Impact scars not evident	1	2	3	4	5	Many scarred trees
Well developed vegetation on bars	1	2	3	4	5	Vegetation regularly stripped from bars
Undisturbed vegetation	1	2	3	4	5	Bent or sprouting trees in overbank area

Total points for vegetation: ____________

Soil and litterfall cover on bars (depth on bars):

In channel thin / mod. / thick
On banks thin / mod. / thick
In flood plain thin / mod. / thick

Evaluation:
Score total: ____________ (in case some categories above are not relevant)
Total scored: ______________
Rating in %: ____________________ (higher % indicates more flooding)

mills outside any village due to their persistence since historic times and exposure against floods.

Frequently, local historians publish their findings in local yearbooks that have to be identified and found in local libraries. Due to missing systematic of both the publications and their content, positive experiences were made by an announcement of the initiation of a project on historic flood evidence and indicators in the regional newspaper. Making such an announcement in the early stages of the investigations and asking the public for support ("citizen science") provide sufficient time for personal visits and contacts afterwards. It is not surprising that the compiled list of hints of floods might contain both gaps and contradictions. Verifications and plausibility checks including neighbouring regions and catchments might become time consuming. Considering the balance between time investment and benefit, the systematic screening of archives makes sense only if catalogues are available or a distinct flood event is already identified, typically by time. Newspaper archives provide remarkable detailed information in such circumstances.

Based on personal experiences, the duration of a historical flood chronology project might take several months depending on availability and access to the sources.

2.6 Methods—Conclusion and Perspectives

As the explanations above have illustrated, the reconstruction of discharge parameters of historical and prehistoric floods is not magic or purely inspirational wishful thinking. The methodological tool is clear, the techniques for estimation comprehensible in all sub-steps. For details, central literature references have been selected that provide further information in an overview and facilitate a more in-depth introduction without getting lost in details. Additionally, in the geomorphological, sedimentological and hydraulic literature there are numerous approaches, not mentioned here, to draw conclusions about flow conditions from structures in the river bed. As indicated above, no exhaustive list of possibilities has been deliberately sought here, but rather the more or less established methods and approaches to flood reconstruction have been concentrated on. Note, that older publications does not mean they are not up to date any more but are chosen as best introduction and review published so far.

In practice, there is seldom the possibility of a large selection of different methods for flood estimation. Rather, one has to evaluate the palaeo-flood indicators that have been handed down, which inevitably reduces the spectrum of methods. Uncertainty analysis is suggested in cases of ranges of values of input data for any approach which provides quantitative information of uncertainty—hopefully not in an overwhelming magnitude. Wohl (1998) carried out an uncertainty analysis for the variations of the estimated roughness coefficient n according to Manning and can reveal that uncertainties in the estimation of the value have less influence than expected. In general, a critical discussion of the resilience and precision of the estimated discharge parameters is essential. Also, taking into account the local conditions of the respective study, one must always be aware of whether minimum, maximum or mean values of the respective parameter are recorded—a thoroughly non-trivial aspect that has often not been carefully considered.

Nevertheless, if possible, several approaches should be taken into account, because this makes it easier to discuss the parameters determined in terms of their robustness and validity. For this purpose, the comparison of results from independent methods is helpful, as some examples might illustrate. For example, the study by Bohorquez and Darby (2008) shows how the plausibility of the results of different hydraulic flow models can be tested using the flow competence, indicated by the grain size of the relocated sediments. Carling and Grodek (1994) discuss the plausibility of discharge values estimated via different approaches (discharge model, competence, channel bed structures) for a single flood in the British Isles. As a consequence, they can show that previous data on floods resulting from maximum area precipitation lead to substantial overestimations. House and Pearthree (1995) evaluated previous estimations of the discharge of a flash due to the exceptional large specific discharge, which draw attention on the flood event rather poorly documented due to the remote location. Herget (2005) documents by the example of reconstructing glacial lake outburst floods in Siberia that by no means all impressively large channel bed structures and flood level indicators can be attributed to the peak discharge, but were created during different stages of the unsteady hydrograph of the repeated flood events. A later by specialists carried out advanced modelling approach for the same event (Bohorquez et al. 2015) allowed chronological sorting of different flood features far away from each other along the flood pathway and provides clues for the interpretation of e.g. gravel dunes that could not be revealed by singles locations separately.

It has been shown above—and hopefully conveyed in a credible manner—that the indirect discharge estimation methods of palaeohydrology can plausibly estimate the magnitudes of various discharge parameters. Although calculations are necessary, the term "estimation" is deliberately used instead of "calculation" in order not to mask the unavoidable imprecision. Note, that algebraically different numbers do not necessarily indicate different results. If the estimation of plausible orders of magnitude is an acceptable result, palaeohydrology can certainly impress with the variety of its methodological approaches and possibilities—the example studies in the following chapters will illustrate this!

References

Acrement GJ, Schneider VR (1989) Guide for selecting Manning's roughness coefficients for natural channels and flood plains. US Geological Survey Water Supply Paper 2339

Alcoverro J, Corominas J, Gómez M (1999) The Barranco de Arás flood of 7 August 1996 (Biescas, Central Pyrenees, Spain). Eng Geol 51:237–255

Alexander J, Cooker MJ (2016) Moving boulders in flash floods and estimating flow conditions using boulders in ancient deposits. Sedimentology 63:1582–1595

Alexandre P (1987) Le climat en Europe au Moyen Age. Paris (École des hautes études en sciences sociales)

Alho P, Aaltonen J (2008) Comparing a 1D hydraulic model with a 2D hydraulic model for the simulation of extreme glacial outburst floods. Hydrol Process 22:1537–1547

Allen JRL (1984) Sedimentary structures—their character and physical basis. Elsevier, Amsterdam

Almeida RP, Freitas BT, Turra BB, Figueiredo FT, Marconato A, Janikian L (2016) Reconstructing fluvial bar surfaces from compound cross-strata and the interpretation of bar accretion direction in large river deposits. Sedimentology 63:609–628

Amsler ML, Schreider MI (1999) Dunes height prediction at floods in the Parana River, Argentina. In: Jayawardena AW, Lee JHW, Wang ZY (eds) River sedimentation. Balkema, Rotterdam, pp 615–620

Apman RP (1973) Estimating discharge from superelevation in bends. J Hydraul Div—Proc Am Soc Civil Eng 99:65–79

Ashley GM (1990) Classification of large-scale subaqueous bedforms—a new look at an old problem. J Sediment Petrol 60:160–172

Baker VR (1973) Paleohydrology and sedimentology of Lake Missoula flooding in eastern Washington. Geol Soc Am Spec Paper 144

Ballesteros JA, Bodoque JM, Diez-Herrero A, Sanchez-Silva M, Stoffel M (2011) Calibration of floodplain roughness and estimation of flood discharge based on tree-ring evidence and hydraulic modelling. J Hydrol 403:103–115

Barben M (2003) Beurteilung von Verfahren zur Abschätzung seltener Hochwasserabflüsse in mesoskaligen Einzugsgebieten. Geographica Bernensia G 71

Barben M, Hodel H-P, Kleeberg H-B, Spreafico M, Weingartner R (2001) Übersicht über Verfahren zur Abschätzung von Hochwasserabflüssen—Erfahrungen aus den Rheinanliegerstaaten. Internationale Kommission für die Hydrologie des Rheingebietes KHR, Bericht I-19

Barnes HH (1967) Roughness characteristics of natural channels. US Geological Survey Water Supply Papers, p 1849

Bathurst JC (1978) Flow resistance of large-scale roughness. J Hydraul Div—Proc Am Soc Civil Eng 104:1587–1603

Beget JE (1986) Comment on "Outburst floods from glacial Lake Missoula" by G.K.C. Clarke, W.H. Mathews & R.T. Pack. Quatern Res 25:136–138

Benito G, Thorndycraft VR (eds) (2004) Systematic, palaeoflood and historical data for the improvement of flood risk estimation—methodological guidelines. Centro de Ciencias Medioambientales, Madrid

Benson MA, Dalrymple T (1967) General field and office procedures for indirect discharge measurements. In: Techniques of water-resources investigations of the United States Geological Survey Book 3, Chapter A1. US Geological Survey, Washington

Bertola M et al (2023) Megafloods in Europe can be anticipated from observations in hydrologically similar catchments. Nat Geosci 16:982–988

Bjerklie DM (2007) Estimating the bankfull velocity and discharge for rivers using remotely sensed river morphology information. J Hydrol 341:144–155

Bohorquez P, Darby SE (2008) The use of one- and two-dimensional hydraulic modelling to reconstruct a glacial outburst flood in a steep Alpine valley. J Hydrol 361:240–261

Bohorquez P, Carling PA, Herget J (2015) Dynamic simulation of catastrophic late Pleistocene glacial-lake drainage, Altai Mountains, central Asia. Int Geol Rev. https://doi.org/10.1080/00206814.2015.1046956

Bomers A, van der Meulen B, Schielen RMJ, Hulscher SJMH (2019) Historic flood reconstruction with the use of an artificial neural network. Water Resour Res 55:9673–9688

Borisova O, Sidorchuk A, Panin A (2006) Palaeohydrology of the Seim River basin, Mid-Russian Upland, based on palaeochannel morphology and palynological data. CATENA 66:53–73

Bradley RW, Venditti JG (2017) Reevaluting dune scaling relations. Earth-Sci Rev 165:356–376

Bürger K, Dostal P, Seidel J, Imbery F, Barriendos M, Mayer H, Glaser R (2006) Hydrometeorological reconstruction of the 1824 flood event in the Neckar River basin (southwest Germany). Hydrol Sci J 51:864–877

Burr DM (2010) Palaeoflood-generating mechanisms on Earth, Mars, and Titan. Global Planet Change 70:5–13

Carling PA (1989) Hydrodynamic models of boulder berm deposition. Geomorphology 2:319–340

Carling PA (1996a) A preliminary palaeohydraulic model applied to late Quaternary gravel dunes—Altai Mountains, Siberia. In: Branson J, Brown AG, Gregory KJ (eds) Global continental changes—the context of palaeohydrology. Geological Society Special Publication, vol 115, pp 165–179

Carling PA (1996b) Morphology, sedimentology and palaeohydraulic significance of large gravel dunes, Altai Mountains, Siberia. Sedimentology 43:647–664

Carling PA (1999) Subaqueous gravel dunes. J Sediment Res 69:534–545

Carling PA, Grodek T (1994) Indirect estimation of ungauged peak discharges in a bedrock channel with reference to design discharge selection. Hydrol Process 8:497–511

Carling PA, Shvidchenko AB (2002) A consideration of the dune: antidune transition in fine gravel. Sedimentology 49:1269–1282

Carling PA, Gölz E, Orr HG, Redecki-Pawlik A (2000) The morphodynamics of fluvial sand dunes in the River Rhine, near Mainz, Germany—I. Sedimentology and morphology. Sedimentology 47:227–252

Carling PA, Bohorquez P, Fan X (2020) Hydraulic control on the development of megaflood runup deposits. Geomorphology 361:107203

Carling PA, Kidson R, Cao Z, Herget J (2003) Palaeohydraulics of extreme flood events—reality and myth. In: Gregory K, Benito G (Hrsg.) Palaeohydrology—understanding global change. Wiley, Chichester, pp 325–336

Carr MH (1996) Water on mars. Oxford University Press, New York

Church MA, McLean DG, Wolcott JF (1987) River bed gravels—sampling and analysis. In: Thorne CR, Bathurst JC, Hey RD (eds) Sediment transport in gravel-bed rivers. Wiley, Chichester, pp 43–88

Clague JJ, Mathews WH (1973) The magnitude of jökulhlaups. J Glaciol 12:501–504

Clarke AO (1996) Estimating probable maximum floods in the upper Santa Ana Basin, southern California, from stream boulder size. Environ Eng Geosci 2:165–182

Comiti F, Lucía A, Rickenmann D (2016) Large wood recruitment and transport during large floods—a review. Geomorphology 269:23–39

Coon WF (1998) Estimation of roughness coefficients for natural stream channels with vegetated banks. US Geological Survey Water-Supply Paper 2441

Costa JE (1983) Paleohydraulic reconstruction of flash-flood peaks from boulder deposits in the Colorado Front Range. Geol Soc Am Bull 94:986–1004

Costa JE (1984) Physical geomorphology of debris flows. In: Costa JE, Fleisher PJ (eds) Developments and applications of geomorphology. Springer, Berlin, pp 268–317

Costa JE (1987) A comparison of the largest rainfall-runoff floods in the Unites States with those of the People's Republic of China and the world. J Hydrol 96:101–115

Costa JE (1988) Floods from dam failures. In: Baker VR, Kochel RC, Patton PC (eds) Flood geomorphology. Wiley, New York, pp 439–463

Costa JE (1988) Rheologic, geomorphic, and sedimentologic differentiation of water floods, hyperconcentrated flows, and debris flows. In: Baker VR, Kochel RC, Patton PC (eds) Flood geomorphology. Wiley, New York, pp 113–122

Crippen JR, Bue CD (1977) Maximum floodflows in the conterminous United States. US Geological Survey Water Supply Paper 1887

Crippen JR (1982) Envelope curves for extreme flood events. J Hydraul Div—Proc Am Soc Civil Eng 108:1208–1212

Crowder DW, Knapp HV (2005) Effective discharge recurrence intervals of Illinois streams. Geomorphology 64:167–184

Cserkész-Nagy Á, Thamó-Bozsó E, Tóth T, Sztanó O (2012) Reconstruction of a Pleistocene meandering river in East Hungary by VHR seismic images and its climatic implications. Geomorphology 153–154:205–218

De Brue H, Poesen J, Notebaert B (2015) What was the transport mode of large boulders in the Campine Plateau and the lower Meuse valley during the mid-Pleistocene? Geomorphology 228:568–578

Desloges JR, Jones DP, Ricker KE (1989) Estimates of peak discharge from the drainage of ice-dammed Ape Lake, British Columbia, Canada. J Glaciol 35:349–354

Deutsch M, Pörtge K-H (2002) Hochwasserereignisse in Thüringen. Schriftenreihe der Thüringer Landesanstalt für Umwelt und Geologie 63

Deutsch M, Glaser R, Pörtge K-H, Börngen M, Drescher A, Martin B, Riemann D, Schönbein J (2010) Historische Hochwasserereignisse in Mitteleuropa—Quellenkunde, Interpretation und Auswertung. Geographische Rundschau 3/2010:18–24

Dury GH (1985) Attainable standards of accuracy in the retrodiction of palaeodischarge from channel dimensions. Earth Surf Proc Land 10:205–213

Eide CH, Müller R, Helland-Hansen W (2018) Using climate to relate water-discharge and area in modern and ancient catchments. Sedimentology 65:1378–1389

Engstrom WN (1996) The California storm of January 1862. Quatern Res 46:141–148

Ethridge FG, Schumm SA (1978) Reconstructing paleochannel morphologic and flow characteristics—methodology, limitations, assessment. Can Soc Petrol Geol Memoir 5:703–721

Euler T, Herget J (2012) Controls on local scour and deposition induced by obstacles in fluvial environments. CATENA 91:35–46

Euler T, Zemke J, Rodrigues S, Herget J (2014) Influence of inclination and permeability of solitary woody riparian plants on local hydraulic and sedimentary processes. Hydrol Process 28:1358–1371

Euler T, Herget J, Schlömer O, Benito G (2017) Hydromorphological processes at solitary boulder obstacles in streams. CATENA 157:250–267

Fasken GB (1963) Guide for selecting roughness coefficient "n" values for channels. US Department of Agriculture—Soil Conservation Service. Lincoln (USDA)

Ferguson RI (2005) Estimating critical stream power for bedload transport calculations in gravel-bed rivers. Geomorphology 70:33–41

Ferguson RI, Sharma BP, Hardy RJ, Hodge RA, Warburton J (2017) Flow resistance and hydraulic geometry in contrasting reaches of a bedrock channel. Water Resour Res 53. https://doi.org/10.1002/2016WR020233

Fernandez R, Best J, López F (2006) Mean flow, turbulence structure, and bed form superimposition across the ripple-dune transition. Water Resour Res 42:W05406. https://doi.org/10.1029/2005WR004330

Fonstad MA, Reichling JP, van de Grift JW (2005) The transparent velocity-head rod for inexpensive and accurate measurement of stream velocities. J Geosci Educ 53:44–52

Foulds SA, Griffiths HM, Macklin MG, Brewer PA (2014) Geomorphological records of extreme floods and their relationship to decadal-scale climate change. Geomorphology 216:193–207

Froude MJ, Alexander J, Barclay J, Cole P (2017) Interpreting flash flood palaeoflow parameters from antidunes and gravel lenses—an example from Montserrat, West Indies. Sedimentology 64:1817–1845

Gábris G (2001) Paléohydrologie tardiglaciaire et holocène en Hongrie d'après l'étude des méandres libres. Revue Geographique De L'est 4(2001):149–156

Glaser R (1996) Data and methods of climatological evaluation in historical climatology. Hist Soc Res 21:56–88

Gottesfeld AS (1996) British Columbia flood scars—maximum flood-stage indicators. Geomorphology 14:319–325

Grant GE (1997) Critical flow constrains flow hydraulics in mobile-bed streams—a new hypothesis. Water Resour Res 33:349–358

Greenbaum N, Schick AP, Baker VR (2000) The Palaeoflood record of a hyperarid catchment, Nahal Zin, Negev Desert, Israel. Earth Surf Proc Land 25:951–971

Greenbaum N, Schwartz U, Carling P, Bergman N, Mushkin A, Zituni R, Halevi R, Benito G, Porat N (2020) Frequency of boulders transport during large floods in hyperarid areas using paleoflood analysis—an example from the Negev Desert, Israel. Earth Sci Rev 202:103086

Gregory KJ (1976) Lichens and the determination of river channel capacity. Earth Surf Proces 1:273–285

Gyllenbok J (2018) Encyclopaedia of historical metrology, weights, and measures (3 vols). Springer, Cham

Harms JC (1969) Hydraulic significance of some sand ripples. Geol Soc Am Bull 80:363–396

Harnischmacher S (2002) Fluvialmorphologische Untersuchungen an kleinen, naturnahen Fließgewässern. Bochumer Geographische Arbeiten 70

Hayden BP (1988) Flood climates. In: Baker VR, Kochel RC, Patton PC (eds.) Flood geomorphology. Wiley, New York, pp 13–26

Herget J (2005) Reconstruction of Pleistocene ice-dammed lake outburst floods in the Altai Mountains. Geological Society of America Special Paper, Siberia, p 386

Herget J, Bremer E, Coch T, Dix A, Eggenstein G, Ewald K (2005) Engineering impact on river channels in the River Rhine catchment. Erdkunde 59 (Heft 3/4):294–319

Herget J, Euler T, Roggenkamp T, Zemke J (2013) Obstacle marks as palaeohydraulic indicators of Pleistocene megafloods. Hydrol Res 44:300–317

Herget J, Roggenkamp T, Krell M (2014) Estimation of peak discharges of historical floods. Hydrol Earth Syst Sci 18:4029–4037

Herschy R (2003) World catalogue of maximum observed floods. IAHS Publication 264, International Association of Hydrological Sciences, Wallingford

Hicks DM, Mason PD (1998) Roughness characteristics of New Zealand Rivers. National Institute of Water and Atmospheric Research, Christchurch

House PK, Pearthree PA (1995) A geomorphologic and hydrologic evaluation of an extraordinary flood discharge estimate—Bronco Creek, Arizona. Water Resour Res 31:3059–3079

Hørland B (2012) Methods for evaluating information sources: an annotated catalogue. J Inf Sci 3/38. https://doi.org/10.1177/0165551512439178

Hulsing H, Smith W, Cobb ED (1966) Velocity-head coefficients and open channels. US Geological Survey Water-Supply Paper 1869-C

Hupp CR, Osterkamp WR (1996) Riparian vegetation and fluvial geomorphic processes. Geomorphology 14:277–295

Ikeda S, Parker G (eds) (1989) River meandering. Water Resources Monograph 12. American Geophysical Union, Washington

Ingram MJ, Underhill DJ, Wigley TML (1978) Historical climatology. Nature 276:329–334

Jackson RG (1976) Sedimentological and fluid-dynamic implications of the turbulent bursting phenomenon in geophysical flows. J Fluid Mech 77:531–560

Jarrett RD, Tomlinson EM (2000) Regional interdisciplinary paleoflood approach to assess extreme flood potential. Water Resour Res 36:2957–2984

Jopling AV (1966) Some principles and techniques used in reconstructing the hydraulic parameters of a paleo-flow regime. J Sediment Petrol 36:5–49

Julien PY, Klaassen GJ (1995) Sand-dune geometry of large rivers during floods. J Hydraul Eng 121:657–663

Kaiser K, Lorenz S, Germer S, Joschus O, Küster M, Libra J, Bens O, Hüttl RF (2012) Late Quaternary evolution of rivers, lakes and peatlands in northeast Germany reflecting past climatic and human impact—an overview. Quatern Sci J 61:103–132

Kalweit H (ed) (1993) Der Rhein unter der Einwirkung des Menschen—Ausbau, Schiffahrt, Wasserwirtschaft. Internationale Kommission für die Hydrologie des Rheingebietes KHR, Bericht I-11

Karcz I (1968) Fluviatile obstacle marks from the wadis of the Negev. J Sediment Petrol 38:1000–1012

Kehew AE, Milewski A, Soliman F (2010) Reconstructing an extreme flood from boulder transport and rainfall-runoff modelling: Wadi Isla, South Sinai, Egypt. Glob Planet Change 70:64–75

Kershaw JA, Clague JJ, Evans SG (2005) Geomorphic and sedimentological signature of a two-phase outburst flood from moraine-dammed Quee Bess Lake, British Columbia, Canada. Earth Surf Proc Land 30:1–25

Kleinhans MG (2004) Sorting in grain flows at the lee side of dunes. Earth Sci Rev 65:75–102

Knox JC (1987) Stratigraphic evidence of large floods in the upper Mississippi Valley. In: Mayer L, Nash D (eds) Catastrophic flooding. Allen & Unwin, Boston, pp 156–180

Kochel RC, Baker VR (1988) Paleoflood analysis using slackwater deposits. In: Baker VR, Kochel RC, Patton PC (Hrsg.) Flood geomorphology. Wiley, New York, pp 357–376

Komar PD (1988) Sediment transport by floods. In: Baker VR, Kochel RC, Patton PC (eds) Flood geomorphology. Wiley, New York, pp 97–111

Komar PD (1989) Flow-competence evaluations of the hydraulic parameters of floods—an assessment of the technique. In: Beven K, Carling P (eds) Floods—hydrological, sedimentological and geomorphological implications. Wiley, Chichester, pp 107–134

Komar PD, Carling PA (1991) Grain sorting in gravel bed-streams and the choice of particle sizes for flow-competence evaluations. Sedimentology 38:489–502

Koutroulis A, Tsanis IK (2010) A method for estimating flash flood peak discharge in a poorly gauged basin—case study for the 13–14 January 1994 flood, Giofiros basin, Crete, Greece. J Hydrol 385:150–164

Lam D, Croke J, Thompson C, Sharma A (2017) Beyond the gorge: palaeoflood reconstruction from slackwater deposits in a range of physiographic settings in subtropical Australia. Geomorphology 292:164–177

Leuschner C, Ellenberg H (2017) Ecology of Central European Forests. Springer, Cham

Li D, Hager WH (1991) Correction coefficients for uniform channel flow. Can J Civ Eng 18:156–158

Limerinos JT (1970) Determination of the Manning coefficient from measured bed roughness in natural channels. US Geological Survey Water-Supply Paper 1898-B

Ljungqvist FC et al (2020) Ranking of tree-ring based hydroclimate reconstructions of the past millennium. Quatern Sci Rev 230:106074

Luterbacher J et al (2010) Circulation dynamics and its influence on European and Mediterranean January–April climate over the past half millennium—results and insights from instrumental data, documentary evidence and coupled climate models. Clim Change 101:201–234

Macklin M, Tooth S, Brewer PA, Noble PL, Duller GAT (2010) Holocene flooding and river development in a Mediterranean steepland catchment—the Anapodaris Gorge, south central Crete, Greece. Global Planet Change 70:35–52

Marcus WA, Roberts K, Harvey L, Tackman G (1992) An evaluation of methods for estimating Manning's n in small mountain streams. Mt Res Dev 12:227–239

Matthes GH (1947) Macroturbulence in natural streams. Am Geophys Union Trans 28(2):255–262

McMahon TA, Finlayson BL, Haines AT, Srikanthan R (1992) Global runoff—continental comparisons of annual flows and peak discharges. Cremlingen (Catena)

Mervalde V (2009) Effect of spatial trends on interpolation of river bathymetry. J Hydrol 371:169–181

Michiels H-G, Aldinger E (2002) Forstliche Standortsgliederung in der badischen Rheinaue. Allgemeine Forstzeitschrift - der Wald 15:811–815

Miller DJ (1960) Giant waves in Lituya Bay, Alaska. US Geological Survey Professional Paper 354-C

Milliman JD, Farnsworth KL (2011) River discharge to the coastal ocean—a global synthesis. Cambridge University Press, Cambridge

Miwa H, Daido A, Kato I (1999) Transformation of sand waves under unsteady conditions in alluvial streams. In: Jayawardena AW, Lee JHW, Wang ZY (eds) River sedimentation. Balkema, Rotterdam, pp 211–216

Müller N (1995) River dynamics and floodplain vegetation and their alterations due to human impact. Archiv Für Hydrobiologie Suppl 101:477–512

Muñoz-Salinas E, Manea VC, Palacios D, Castillo-Rodriguez M (2007) Estimation of lahar flow velocity on Popocatépetl volcano, Mexico. Geomorphology 92:91–99

Nadolny I (1994) Morphologie und Hydrologie naturnaher Flachlandbäche unter gewässertypologischen Gesichtspunkten. Universität Karlsruhe, Mitteilungen des Institutes für Wasserbau und Kulturtechnik, p 189

Naudascher E (1992) Hydraulik der Gerinne und Gerinnebauwerke. Springer, Wien and New York

Ng F, Björnsson H (2003) On the Clague-Mathews relation for jökulhlaups. J Glaciol 49:161–172

Nicoll TJ, Hickin EJ (2010) Planform geometry and channel migration of confined meandering rivers on the Canadian prairies. Geomorphology 116:37–47

Ochs K, Egger G, Weber A, Ferreira T, Householder JE, Schneider M (2020) The potential natural vegetation of large river floodplains—from dynamic to static equilibrium. J Hydro-Environ Res 30:71–81

Osterkamp WR, Hedman ER (1982) Perennial-streamflow characteristics related to channel geometry and sediment in Missouri River basin. US Geological Survey Professional Papers 1242

Padgett GV, Ehrlich R (1976) Paleohydrologic analysis of a late Carboniferous fluvial system, southern Morocco. Geol Soc Am Bull 87:1101–1104

Park CC (1977) Man-induced changes in stream channel capacity. In: Gregory KJ (ed) River channel changes. Wiley, Chichester, pp 121–144

Park CC (1977b) World-wide variations in hydraulic geometry exponents of stream channels—an analysis and some observations. J Hydrol 33:133–146

Pauling A, Luterbacher J, Casty C, Wanner H (2006) Five hundred years of gridded high-resolution precipitation reconstructions over Europe and the connection to large-scale circulation. Clim Dyn 26:387–405

Pierson TC (1985) Initiation and flow behaviour of the 1980 Pine creek and Muddy River Lahars, Mount St. Helens, Washington. Geol Soc Am Bull 96:1056–1069

Pierson TC (2005) Hyperconcentrated flow—transitional process between water flow and debris flow. In: Jakob M, Hungr O (eds) Debris-flow hazards and related phenomena. Springer, Berlin, 159–202

Potter KW, Walker JF (1985) An empirical study of flood measurement error. Water Resour Res 21:403–406

Porter K et al (2011) Overview of the ARkStorm scenario. US Geological Survey Open-File Report 2010–1312
Pruess J, Wohl EE, Jarrett RD (1998) Methodology and implications of maximum paleodischarge estimates for mountain channels, upper Animas River catchment, Colorado, USA. Arct Alp Res 30:40–50
Quick MC (1991) Reliability of flood discharge estimates. Can J Civ Eng 18:624–630
Reist T, Weingartner R, Gurtz J (2002) Neue Wege bei der Beschreibung alter Hochwasser—die Wassernot im Emmental 13. August 1837. Wasser & Boden 54/10:50–54
Roggenkamp T, Herget J (2016) Middle and Lower Rhine in Roman times—a reconstruction of hydrological data based on historical sources. Environ Earth Sci 75:1–12
Roggenkamp T, Herget J (in review) River Ahr flood of July 2021—a rather frequent exceptional event. Global and Planetary Change
Roggenkamp T (2023) Brückenbauten als Spiegel vergangener Hydrologie. Geogr Rundsch 5(2023):26–30
Rotnicki K (1991) Retrodiction of palaeodischarge of meandering and sinous alluvial rivers and its palaeohydroclimatic implications. In: Starkel L, Gregory KJ, Thornes JB (eds) Temperate palaeohydrology. Wiley, Chichester, pp 431–471
Ruiz-Villanueva V et al (2019) Characterization of wood-laden flows in rivers. Earth Surf Proc Land 44:1694–1709
Saharia M, Kirstetter P-E, Vergara H, Gourley JJ, Hong Y (2017) Characterization of floods in the United States. J Hydrol 548:524–535
Schlömer O, Herget J, Euler T (2020) Boundary condition control of fluvial obstacle mark formation—framework from a geoscientific perspective. Earth Surf Proc Land 45:189–206
Shilpakar RL, Thomas MC, Reid MA (2021) The resilience of a floodplain vegetation landscape. Landscape Ecol 36:139–157
Sidorchuk A, Borisova O, Panin A (2001) Fluvial response to the late Valdai/Holocene environmental change on the East European Plain. Global Planet Change 28:303–318
Sidorchuk AY, Borisova OK (2000) Method of paleogeographical analogues in paleohydrological reconstructions. Quatern Int 72:95–106
Slattery MC (2015) Barbara A Kennedy (1972)—"Bankfull" discharge and meander forms. Prog Phys Geogr 39:275–278
Smith RF, Boardman J (1989) The use of soil information in the assessment of the incidence and magnitude of historical flood events in upland Britain. Beven K, Carling P (eds) Floods—hydrological, sedimentological and geomorphological implications. Wiley, Chichester, pp 185–197
Smith MW, Cox NJ, Bracken LJ (2007) Applying flow resistance equations to overland flow. Prog Phys Geogr 31:363–387
Starkel L (ed) (1996) Evolution of the Vistula river valley during the last 15000 years—Part VI. Polish Academy of Sciences—Geographical Studies, Special Issue 9
Sudhaus D, Seidel J, Bürger K, Dostal P, Imbery F, Mayer H, Glaser R, Konold W (2008) Discharges of past flood events based on historical river profiles. Hydrol Earth Syst Sci Discuss 5:323–344
Sundborg Å, Calles B (2001) Water discharges determined from sediment distributions—a palaeohydrological method. Geogr Ann A83:39–54
Tabasum T, Bhat P, Kumar R, Fatma T, Trisal CL (2009) Vegetation of the river Yamuna floodplain in the Delhi stretch, with reference to hydrological characteristics. Ecohydrology 2:156–163
Tetzlaff G, Börngen M, Mudelsee M, Raabe A (2002) Das Jahrtausendhochwasser von 1342 am Main aus meteorologisch-hydrologischer Sicht. Wasser & Boden 54(10):41–49
Thornes JB (2003) Palaeohydrological modelling—from palaeohydraulics to palaeohydrology. In: Gregory K, Benito G (eds) Palaeohydrology—understanding global change. Wiley, Chichester, pp 291–305
Thorne SD, Furbish DJ (1995) Influences of coarse bank roughness on flow within a sharply curved river bend. Geomorphology 12:241–257
Timár G, Gábris G (2008) Estimation of water conductivity of the natural flood channels on the Tisza flood-plain, the Great Hungarian Plain. Geomorphology 98:250–261
Toonen WHJ (2013) A Holocene flood record of the River Rhine. Utrecht Studies in Earth Sciences, vol 41
Toonen WHJ, Winkels TG, Cohen KM, Prins MA, Middelkoop H (2015) Lower Rhine historical flood magnitudes of the last 450 years reproduced from grain-size measurements of flood deposits using End Member Modelling. CATENA 130:69–81
Toonen WHJ, Munoz SE, Cohen KM, Macklin MG (2020) High-resolution sedimentary paleoflood records in alluvial river environments—a review of recent methodological advances and application to flood hazard assessment. In: Herget J, Fontana A (eds) Palaeohydrology—traces, tracks and trails of extreme events. Springer, Cham, pp 213–228
Turowski JM, Rickenmann D, Dadson SJ (2010) The partitioning of the total sediment load of a river into suspended load and bedload—a review of empirical data. Sedimentology 57:1126–1146
van der Meulen B (2021a) Historic river morphology. Utrecht Stud Earth Sci 236:73–87
van der Meulen B, Defilet MP, Tebbens LA, Cohen KM (2022) Palaeoflood level reconstructions in a lowland setting from urban archaeological stratigraphy, Rhine river delta, the Netherlands. CATENA 212:106031
Vandenberghe J (2003) Climate forcing of fluvial system development—an evolution of ideas. Quatern Sci Rev 22:2053–2060
Vandenberghe J, Sidorchuk A (2020) Large palaeomeanders in Europe—distribution, formation process, age, environments and significance. In: Herget J, Fontana A (eds) Palaeohydrology. Switzerland, Cham, pp 169–186
Vogel RM, Zafirakou-Koulouris A, Matalas NC (2001) Frequency of record-breaking floods in the United States. Water Resour Res 37:1723–1731
Walder JS, Costa JE (1996) Outburst floods from glacier-dammed lakes—the effect of mode of lake drainage on flood magnitude. Earth Surf Proc Land 21:701–723
Weikinn C (1963) Quellentexte zur Witterungsgeschichte Europas von der Zeitenwende bis zum Jahre 1850 - Band 4, 1701 bis 1750. Akademie, Berlin
Wharton G (1992) Flood estimation from channel size—guidelines for using the channel-geometry method. Appl Geogr 12:339–359
Wharton G (1995a) Information from channel geometry-discharge relations. In: Gurnell A, Petts G (eds) Changing river channels. Wiley, Chichester, pp 325–346
Wharton G (1995b) The channel-geometry method—guidelines and applications. Earth Surf Proc Land 20:649–660
Wieczorek GF, Larsen MC, Eaton LS, Morgan BA, Blair JL (2001) Debris-flow and flooding hazards associated with the December 1999 storm in coastal Venezuela and strategies for mitigation. US Geological Survey Open File Report 01–0144
Wilbers A (2004) The development and hydraulic roughness of subaqueous dunes. Netherlands Geographical Studies, vol 323
Williams GP (1978) Bank-full discharge of rivers. Water Resour Res 14:1141–1154

Williams GP (1983) Paleohydrological methods and some examples from Swedish fluvial environments I—cobble and boulder deposits. Geografiska Annaler 65A:227–243

Williams GP (1984a) Paleohydrologic equations for rivers. In: Costa JE, Fleisher PJ (eds) Developments and applications of geomorphology. Springer, Berlin, pp 343–367

Williams GP (1984b) Paleohydrological methods and some examples from Swedish fluvial environments II—river meanders. Geografiska Annaler 66A:89–102

Williams GP (1986) River meanders and channel size. J Hydrol 88:147–164

Williams GP, JE Costa (1988) Geomorphic measurements after a flood. In: Baker VR, Kochel RC, Patton PC (eds) Flood geomorphology. Wiley, New York, pp 65–77

Wiman C, Harden T, Shen Z, Curry BB, Reinders JB, Beighley RE, Munoz SE (2023) Large floods on the lower Ohio River inferred from slackwater deposits. Prog Phys Geogr 48:79–95

Wohl EE (1998) Uncertainty in flood estimates associated with roughness coefficients. J Hydraul Eng 124:219–223

Wohl EE, Wilcox A (2005) Channel geometry of mountain streams in New Zealand. J Hydrol 300:252–266

Wolman MG, Costa JE (1984) Envelope curves for extreme flood events. J Hydraul Eng 110:77–78

Wundt W (1965) Grenzwerte der Hochwasserspende und der mittleren Abflussspende in Abhängigkeit von der Fläche. Die Wasserwirtschaft 55(1):1–4

Yanosky TM, Jarrett RD (2002) Dendrochronologic evidence for the frequency and magnitude of paleofloods. In: House PK, Webb RH, Baker VR, Lavish DR (eds) Ancient floods, modern hazards—principles and applications of paleoflood hydrology. Water Science and Application 5. American Geophysical Union, Washington, pp 77–89

Zanke U (1982) Grundlagen der Sedimentbewegung. Springer, Berlin

Further Reading

There are numerous examples of anthropogenic river channel modifications analysed and documented from all over the world. Even though already published several years ago, the review edited by Ken Gregory provides a systematic review on relevant processes and effects in general:

Gregory KJ (ed) (1977) River channel changes. Wiley, Chichester

In addition to publications on historic climatology, a review on sources of information on historic floods in general and on flood marks on buildings including their analysis and interpretation is presented by:

Barriendos M, Coeur D (2004) Flood data reconstruction in historical times from non-instrumental sources in Spain and France. In: Benito G, Thorndycraft VR (eds) Systematic, palaeoflood and historical data for the improvement of flood risk estimation—methodological guidelines. Centro de Ciencias Medioambientales, Madrid, pp 29–42

van der Meulen B (2021b) Flood marks on buildings. Utrecht Stud Earth Sci 236:45–71

Archaeological evidence of floods and especially consideration of flood levels by water engineering projects including bridges is compiled in review in:

Brown AG (1997) Alluvial geoarchaeology—floodplain archaeology and environmental change. Cambridge University Press, Cambridge

Prell M (ed) (2011) Archäologie der Brücken—archaeology of bridges. Pustet, Regensburg

Violett P-L (2007) Water engineering in ancient civilizations. International Association of Hydraulic Engineering and Research, Madrid

Natural floodlevel indicators and their interpretation is presented with varying focus by:

Baker VR, Kochel RC (1988) Flood sedimentation in bedrock fluvial channels. In: Baker VR, Kochel RC, Patton PC (eds) Flood geomorphology. Wiley, New York, pp 123–137

Ballesteros-Cánovas JA, Stoffel M, St George S, Hirschboek K (2015) A review on flood records from tree rings. Prog Phys Geogr 39:794–816

Bridge JS (2003) Rivers and floodplains—forms, processes, and sedimentary record. Blackwell, Oxford

Herget J (2020) Palaeostage indicators in rivers—an illustrated review. In: Herget J, Fontana A (eds) Palaeohydrology—traces, tracks and trails of extreme events. Springer, Cham, pp 187–211

Jarrett RD, England JF (2002) Reliability of paleostage indicators for paleoflood studies. In: House PK, Webb RH, Baker VR, Lavish DR (eds) Ancient floods, modern hazards—principles and applications of paleoflood hydrology. Water Science and Application 5, American Geophysical Union, Washington, pp 91–109

Wilhelm B et al (2018) Interpreting historical, botanical, and geological evidence to aid preparations for future floods. Wires Water 2018:e1318

Rather easy to read introductions to hydraulics of flowing water and several aspects related to the Manning equation including validity, history and parameter derivation are compiled by:

Chanson H (1999) The hydraulics of open channel flow. Arnold, London

Chow VT (1959) Open channel hydraulics. McGraw-Hill, New York

French RH (1994) Open-channel hydraulics. McGraw-Hill, Singapore

Henderson FM (1966) Open channel flow. Macmillan, New York

Yen BC (ed) (1992) Channel flow resistance—centennial of Manning's formula. Water Resources Publications, Littleton

Reviews and introductions on sediment transport and deposition considering involved processes relevant for hydraulic interpretation are presented by:

Allen JRL (1965) A review of the origin and characteristics of recent alluvial deposits. Sedimentology 5:89–191

Bridge JS (2003) Rivers and floodplains—forms, processes, and sedimentary record. Blackwell, Oxford

Brown AG (1997) Alluvial geoarchaeology—floodplain archaeology and environmental change. Cambridge University Press, Cambridge

Anderson MG, Walling DE, Bates PD (eds) (1996) Floodplain processes. Wiley, Chichester

Reviews on bedforms in river channels including consideration of related generation processes are provided by:

Allen JRL (1984) Sedimentary structures—their character and physical basis. Elsevier, Amsterdam

Reineck H-E, Singh IB (1980) Depositional sedimentary environments. Springer, Berlin

Richardson K, Carling PA (2005) A typology of sculpted forms in open bedrock channels. Geological Society of America Special Paper 392

While reference books on palaeohydrological discharge modelling are missing so far, commented reviews on the development and benchmark papers on the conceptional and methodological development of the discipline might provide a suitable review:

Baker VR (2014) Palaeohydrology—benchmark papers in hydrology. International Association of Hydrological Sciences, Wallingford, vol 9

Thornes JB (2003) Palaeohydrological modelling—from palaeohydraulics to palaeohydrology. In: Gregory K, Benito G (eds) Palaeohydrology—understanding global change. Wiley, Chichester, pp 291–305

Reviews and numerous references to additional studies on discharge estimations and related parameters by remote sensing is provided e.g. by

King TV, Neilson BT, Rasmussen MT (2018) Estimating discharge in low-order rivers with high-resolution aerial imagery. Water Resour Res 54:863–878. https://doi.org/10.1002/2017WR021868

Oubanas H, Gejadze I, Malaterre P-O, Durand M, Wei R, Frasson RPM, Domeneghetti A (2018) Discharge estimation in ungauged basins through variational data assimilation—the potential of the SWOT Mission. Water Resour Res 54:2405–2423. https://doi.org/10.1002/2017WR021735

Pavelsky TM (2014) Using width-based rating curves from spatially discontinuous satellite imagery to monitor river discharge. Hydrol Process 28:3035–3040. https://doi.org/10.1002/hyp.10157

3 Floods in Historic Times

Insofar as written records have been compiled, one can find records of floods from historic times from all parts of the world (cf. Table 3.1). Often, these descriptions are related to weather records, namely for extraordinary events. As expected, the oldest flood records are from the ancient civilisations in China and Egypt and thus date back to the time before the birth of Christ.

As explained above, the flood descriptions that have been handed down still require source-critical analysis in order to be able to assess the resilience and reliability of the information. One must also be aware that in many cases only subjective qualitative descriptions or damage information has been given, not hydrological data or even water levels. Nevertheless, this information forms pieces of the mosaic of an overall picture that will be outlined below.

The aim here is deliberately not to provide a catalogue of all floods worldwide—such compilations, which also includes isolated data on floods before the start of regular measurements, are already available (e.g. Rodier and Roche 1984, Herschy 2003, Milliman and Farnsworth 2011, Global Runoff Data Centre www.grdc.bafg.de). Instead, this overview will initially present flood chronologies of selected catchments as examples. Exceptional events are presented below and interregional correlations are made. An outlook on flood frequencies rounds off the picture, before the example of a historic flood of the River Rhine in Cologne is used to illustrate a transparent approach to estimate the discharge of historical floods. In Table 3.1, central compilations of historical source texts, data lists and regional flood chronologies are listed to complement the examples selected here.

3.1 Historical Floods in Selected Regions and Catchments

For illustration how historic flood chronologies look like and which background information might become relevant, results of studies of flood conditions in historic times are presented below. The selection of several countries is partly arbitrary but also influenced by personal experiences of the author and interesting regional boundary conditions of flood dynamics. A focus is given to put recent prominent flood events into historic perspective.

3.1.1 Germany

- **River Rhine**

As Germany's largest river, the River Rhine has repeatedly demanded and received special attention with regard to its floods. Well remembered are the two floods of the century in short succession in 1993 and 1995 (Engel 1999). In the middle and lower reaches of the river, these events are among the largest within the measurement period (Fig. 3.1).

However, these floods were not only out of the bounds of experience due to the peak discharges: the rise of the flood wave in Cochem on the prominent tributary River Moselle, for example, showed a previously unknown steepness of 5 m in 24 h, which required increased efforts for flood protection. After the analysis of flood generation and dynamics (Engel 1999), flood protection measures in Cologne, for example, were better adapted to the discharges, but also the question of the flood maxima to be expected in the future became topical again. Against the background of ongoing climate change, forecasts were made (Grabs 1997; Görgen 2010), the effects of river engineering measures and retention areas were examined (Lammersen et al. 2002), but also a systematic look at weather and floods in the past was taken (Witte et al. 1995; Krahe 1997; Krahe and Larina 2010). This illustrates that the maxima of the water level records of the gauge have been reached and even exceeded several times before (Fig. 3.2).

A distinction must be made here between winter, summer and ice-jam floods. Floods in the winter half-year are caused by rain, snowmelt or a combination of both.

J. Herget, *Flood Reconstruction*, Springer Natural Hazards, https://doi.org/10.1007/978-3-031-66815-9_3

Table 3.1 Compilations of source texts, data and regional chronologies of historical floods (supplemented, updated and modified considering Brázdil et al. (2006), Herget (2012) and Kiss (2019))

(A) Source text compilations: • Europe: Henning 1904, Weikinn (1958, 1960, 1961, 1963), Börngen and Tetzlaff (2000, 2002) • Northern Germany: Kuß (1825) • Scotland: McEwen (1987) Like the list by Henning (1904), the six volumes of the historical weather conditions description texts edited by Weikinn and completed by Börngen and Tetzlaff contain numerous contemporary references for the period from the turn of the eras up to AD 1850. Note, that the individual historic references still require a source-critical analysis respectively a verification by alternative sources
(B) Data bases: While the first group of databases focus on historical floods and droughts respectively hydrological data including historic times • CBHS Chronology of British Hydrological Events—a public repository for hydrological facts of the type that come from texts rather than tables: https://cbhe.hydrology.org.uk • Repères de Crues – database and web map application of > 70,000 floodmarks in France: https://www.reperesdecrues.developpement-durable.gouv.fr/ • ORRION Observatoire Régional des Risques d'inondation – database on historical floods in Alsace and Upper Rhein Valley: https://orrion.fr/ • Historical floods MAP – compilation and map application of historical flood data for Czech Republic: https://chmi.maps.arcgis.com • GRDC Global Runoff Data Centre—an international archive of data up to 200 years old, fosters multinational and global long-term hydrological studies: http://grdc.bafg.de • UNDINE – information platform on hydrological extreme events covering the major rivers of Germany and combining historical with recent data with a focus on water quality during extreme events: https://undine.bafg.de/ the second group documents data related to historical climatology that includes flood events: • TAMBORA the climate and environmental history collaborative research environment—a focus is given on meteorological information and historical climatology: http://www.tambora.org • Euro-Climhist—evidence on weather and climate including daily weather observations as well as extreme events and long-term climatic development in Switzerland: https://www.euroclimhist.unibe.ch/ All data-bases are accessible without charge and many can be read in different languages
(C) Data tables on flood events: • Europe in Medieval times: Alexandre (1987) • Discharge data from around the world, including a significant amount of historic data: Parde (1961), Rodier and Roche (1984), Francou and Rodier (1967), Herschy (2003), Miliman and Farnsworth (2011)
(D) Regional flood chronologies: The following compilation of references on flood chronologies focusing on historic times does not consider local studies and investigations of distinct flood events; reference on these are given in the listed publications. Further details on selected regions are given in the following subchapters. Note, that this list is far from being complete and the spatial sorting varies in the compiled references **Global or continental scaled reviews:** • Examples and review from around the world: O'Connor et al. (2002), O'Connor and Costa (2004), Parde (1961), Snorasson et al. (2002), Thorndycraft et al. (2003) • Europe: Brázdil et al. (1999, 2012), Glaser and Stangl (2004), Glaser (2008), Glaser et al. (2010) **Egypt:** • Bell (1970) **Namibia:** • Grodek et al. (2013) **India:** • Kale (1998) **China:** • Fucheng et al. (1987), Luo et al. (1987), Pang et al. (1987) **Fiji-Islands:** • Yeo et al. (2007) **USA:** • Reviews USA: Costa and Jarrett (2008), Hoyt and Langbein (1955), O'Connor and Costa (2003) • Mississippi: Munoz et al. (2018) **Mexico:** • Valdés-Manzanilla (2016) **Argentina:** • Prieto (2007)

(continued)

Table 3.1 (continued)

Germany:
• Review Germany: Henning 1904, Schmidt (2000), Sturm et al. (2001a, b)
• Rivers Lech and Isar: Böhm and Wetzel (2006)
• River Rhine: Gerhard (1985), Himmelsbach et al. (2015), Schmidt (2002), Weber (1977), Witte et al. (1995)
• River Werra: Deutsch et al. (2004), Mudelsee et al. (2006)
• River Main: Glaser (1998)
• River Neckar: Röckel (1994)
• River Mosel: Sartor (2012)
• River Elbe: Pötzsch (1784, 1786)
• Thuringia: Deutsch and Pörtge (2002, 2009)
• Saxony: Fügner (1995), Pohl (2004a, b)
• Baden-Württemberg: LUMN (2006)
• Bavaria: Böhm (2011), Böhm et al. (2015), LfU (2008)
Belgium / Netherlands:
• Buisman and van Engelen (1995, 2006), Gottschalk (1971, 1977), Tol and Langen (2000), Glaser and Stangl (2003), De Kraker (2006), Toonen (2015), van der Meulen (2021)
France:
• Review France: Champion (1858–1864), Parde (1961), Lang and Cœur (2014)
• Normandy: Agasse (2003)
• Alsace: Descombes (1995)
• River Ardèche: Naulet et al. (2005)
• River Loire: Dion (1961), Duband (1996)
• River Rhone (French part): Pichard and Roucaute (2014), Pichard et al. (2017)
• River Doubs: Blazer (1931)
• River Isère: Cœur (2008)
United Kingdom:
• Review United Kingdom: Jones et al. (1984), Williams and Archer (2002), Macdonald and Sangster (2017)
• River Tay: Macdonald et al. (2006)
• River Ouse: Macdonald and Black (2010)
• River Trent: Macdonald (2013)
Scandinavia:
• Norway: Roald (2021)
• Sweden: Retsö (2015)
Austria, Switzerland:
• Switzerland: Gees (1997), Pfister and Hächler (1991), Pfister (1999), Pfister and Wanner (2021), Röthlisberger (1991), Schulte et al. (2015), Vischer (2003)
• River Rhine (Swiss part): Pfister (2006), Wetter et al. (2011)
• Eastern Alps: Rohr (2007)
• River Danube (Austrian part): Kresser (1957), Rohr (2005)
• Rivers Salzach and Inn: Rohr (2003)
• River Traun: Rohr (2010)
Spain:
• Review Spain: Barriendos and Martin-Vide (1998), Barriendos and Rodrigo (2006)
• Catalonia: Llasat et al. (2005), Barrera et al. (2006), Barriendos et al. (2014), Barrera and Llasat (2015)
• SE Spain: Machado et al. (2011)
• River Tagus: Benito et al. (2003)
Italy:
• Review Italy: Pavese et al. (1992)
• River Arno: Caporali et al. (2005)
• River Po: Camuffo and Enzi (1996), Guidoboni (1998)
• River Tiber: Camuffo and Enzi (1996), Schneider (2001), Alessandroni and Remedia (2002), Calenda et al. (2005), Aldrete (2006)
Eastern Europe:
• Czech Republik: Brázdil et al. (2005)
• Rivers Vltava and Elbe (Czech part): Brázdil (1998), Elleder (2015)
• Poland: Ghazi et al. (2023)
• River Vistula: Cyberski et al. (2006)
• Hungary: Kiss (2019)
• Slovakia: Horváthová (2003)
• River Danube (Slovakian part): Svoboda et al. (2000), Pekárová et al. (2013)
• Russia: Georgiadi (1993)

Fig. 3.1 Flood marks of the River Rhine at the ferry in Königswinter-Niederdollendorf. From top to bottom the following water levels are marked: 1845–1993-1 Jan. 1926–1920–1948–1882–1970–1955–1958–1980

Summer floods (in the Cologne area) are caused exclusively by heavy and widespread precipitation in the catchment area. Ice-jam floods can occur when the ice cover of the frozen river course breaks open, the ice floes build up in front of obstacles and bottlenecks and thereby dam up the river course (Beltaos 2008). As a result, extreme water levels are reached without being caused by increased runoff (cf. the flood of 1784 in Sect. 3.2.), which makes direct comparison with the other events difficult.

Ice-jam floods no longer occur in recent times due to the warming of the river water as a result of the discharge of cooling water and waste water. Due to the backwater, however, they reached water levels in 1784 that were not exceeded in historical times. As expected, the winter floods outweigh the summer floods in terms of their number. However, the frequency of the floods is subject to change. Assuming that from the eleventh century onwards all extreme floods and from the fourteenth century onwards all severe floods have been recorded, Krahe and Larina (2010) present the following trends, which are also illustrated in Fig. 3.3: Increase in the frequency of floods from the second half of the fourteenth century, more frequent occurrence of summer floods in earlier periods, namely 1250–1350 and 1400–1500, and currently tending to increase flood frequency.

There are several explanations for the significantly lower number and frequency of floods in the period before 1350, which still require further analysis. For example, the damage potential caused by floods may have been lower due

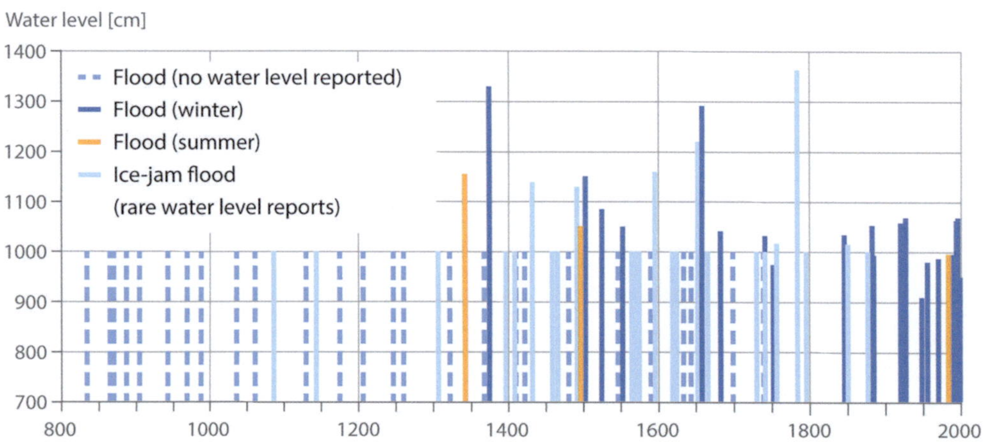

Fig. 3.2 Extreme flood events in Cologne since the Middle Ages (based on data from Krahe and Larina 2010, supplemented). Floods with water levels > 950 cm above the current gauge datum at Cologne (34.972 m above sea level) are differentiated according to summer, winter and ice floods. The latter are not hydraulically comparable in terms of discharge due to the backwater during ice-jams, but result in extreme water levels. In the absence of records of water levels, a value of 900 cm was set as a temporal marker for summer and winter floods, and 1000 cm for ice-jams. Note, that not all floods are not shown in the figure—hence, it should not be interpreted for flood frequencies

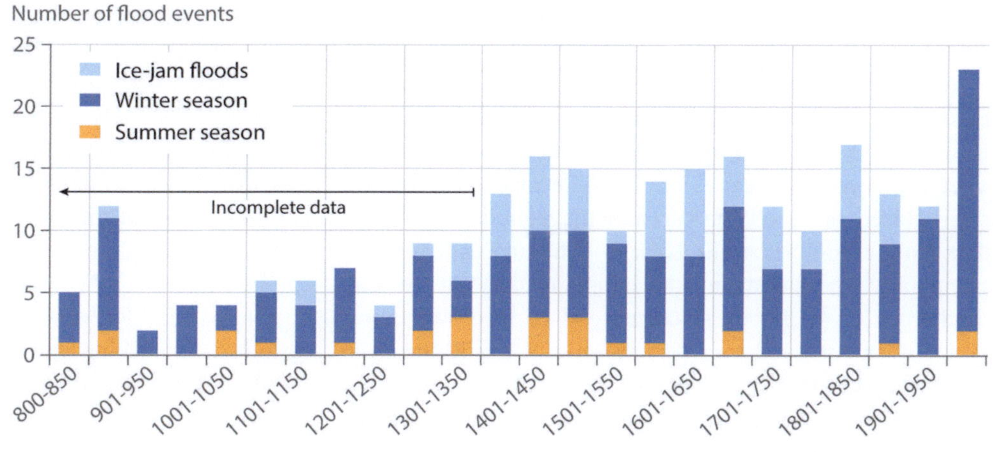

Fig. 3.3 Frequency of severe and extreme floods in Cologne for fifty-year periods since AD 800 (Krahe and Larina 2010)

to the lower population density, so that they are underestimated in terms of magnitude and frequency. Note, that typically only damages but not quantitative data on the flood magnitude are reported. It should also be considered that less intensive land use may have reduced flood formation and intensity. Of course, the floods may actually have been fewer and weaker, but the probability of artefacts should be kept in mind.

In order to be able to better assess the indicated trends and tendencies, a comparison with other river sections and catchment areas is useful. However, it should be noted in advance that the discharge regime of the River Rhine, like that of its tributaries, changes from the upper to the lower course and therefore flood formation in the High and Upper Rhine is largely decoupled from that in the Middle and Lower Rhine (Belz et al. 2007). This is due to the frequent occurrence of floods in the tributaries in the middle course of the Rhine, such as the Rivers Neckar, Main and Moselle, which naturally cannot affect the runoff in the upper course. On the other hand, these tributaries also have a dampening effect on runoff peaks from the Alpine tributaries, which can be quantitatively relativized by the mean flow of the tributaries (Fig. 3.4).

In contrast, it is not surprising that analyses of historical floods in the Dutch Rhine delta show a pattern corresponding to the Lower Rhine (Glaser and Stangl 2003; van der Meulen 2021). Apart from individual cases, the inflows into the Rhine in its lower course do not change the discharge pattern and thus ultimately also the flood history in the long run. However, intensive dike construction as well as diversions of the river courses in historical times further complicate the analysis of the historic flood events (Tol and

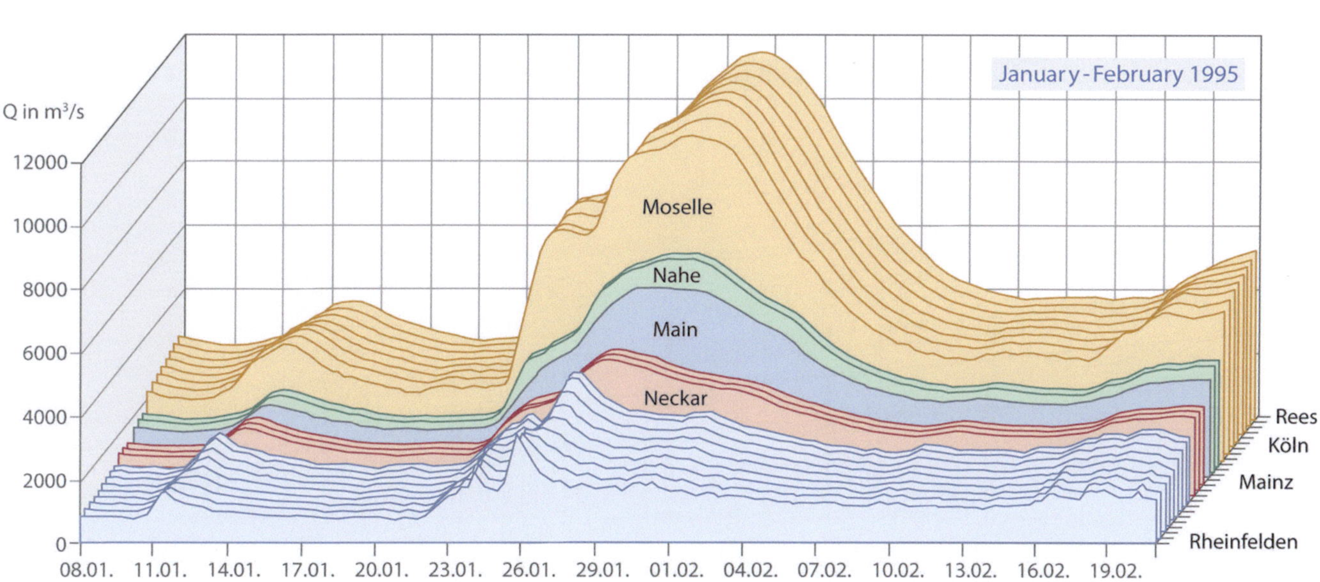

Fig. 3.4 Course of the Rhine flood wave in January/February 1995 (Engel 1999). The figure illustrates by way of example how the floods in the middle and lower reaches of the Rhine are caused by the flood-bearing tributaries, namely the Main and Moselle. No strong floods were observed on the High and Upper Rhine

Langen 2000; Pierik et al. 2018) and complicate the supraregional correlation.

In contrast, the flood history of the High Rhine above Basel looks completely different. In an overview Wetter et al. (2011) also investigate the influence of extensive river engineering measures on floods in this section of the river. Based on flood marks from the period 1641–1882, data from chronicles going back to 1268 and calibrated on comparative measurements, the authors reconstruct the peak discharges of 49 floods since the end of medieval times. Based on a surveyed transverse profile of the Rhine from 1819, the quantitative flood level data can be transferred and related to other water level descriptions. Since numerous buildings are still preserved today, qualitative water level descriptions such as "*Boats had to be loaded through the windows of the guild house of the shipmen*" or "*People on the bridge washed their hands in the Rhine*" can also be quantified and thus associated water levels estimated. On the basis of a wealth of information on flood levels within the urban area of Basel, gradual gradations of the water level can thus be carried out depending on the accuracy and reliability of the source text. In the case of extreme flood levels, it is understandable that the historical sources expressed more their horror and impression than differentiated water level descriptions down to the centimetre. In the absence of further differentiation, the six extreme floods of 1268, 1275, 1302, 1404, 1451 and 1679 have an identical maximum level. From the available elevation data of the floodplain of the Rhine in Basel, the cross-section area can be calculated. Together with estimates of the flow velocity—for this the model FLUX/FLORIS^{2000} was used, unfortunately more detailed information is missing—the discharge can now be calculated (Fig. 3.5).

As expected, the share of floods in the summer months June to August is generally highest with 52% of all severe floods, while winter (21%), spring (16%) and autumn floods (11%) are significantly less frequent. Here, the above-mentioned different discharge regime of the river is characteristically noticeable—floods from catchments in the German mountain ranges reshape the flood pattern along the River Rhine. Even if not all but only severe floods are considered in the evaluation presented here, the period 1651–1700 stands out with a high flood frequency, which can be explained by a related wetter period.

The period 1877–1999, in which no severe floods were observed, is striking and in need of explanation. Extreme floods of catastrophic dimensions at Basel have only been recorded until 1679 and have not been observed since. In addition to hydro-climatic changes, river diversions in the river catchment area deserve special attention in this context (Vischer 2003). As early as 1714, the River Kander was redirected from its original inflow into the River Aare, the most important tributary of the Rhine above Basel, into Lake Thun. Due to the well-known strong flood-buffering effect of lakes, a causal connection could be seen not least through the timing of the absence of extreme floods. The diversion of the River Aare itself through Lake Biel in 1877 also correlates with the beginning of the temporal gap in the flood record. In order to test the hypothesis of the lasting influence of these river diversions on flood reduction in the High Rhine, Wetter et al. (2011) carried out modelling of recent floods, whereby the diversions into the lakes as well as meander culverts and impoundments by hydropower plants carried out in the meantime were not taken into account. The comparison between the floods modelled in this way and the actual measured values shows that the peak discharges have been noticeably reduced and the arrival of the flood waves delayed. However, it should also be noted that the floods were obviously also reduced hydroclimatically during this period, because the water level of

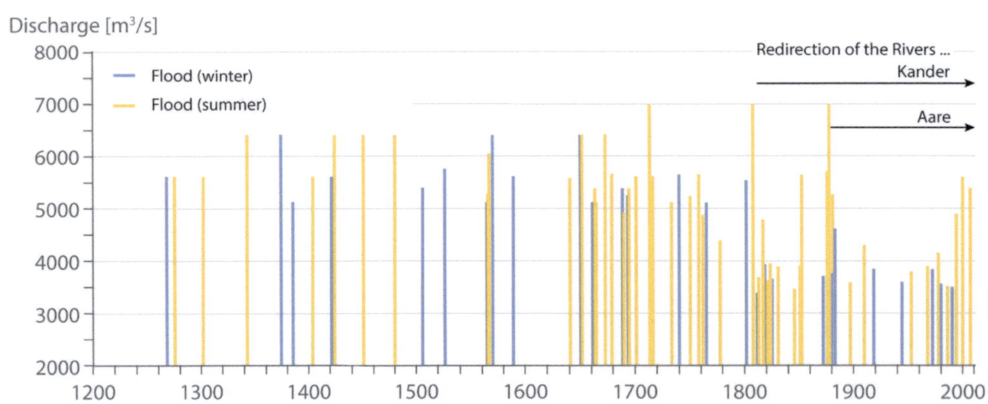

Fig. 3.5 Peak discharges of historical floods of the Rhine in Basel 1250–2010 (data from Wetter et al. 2011). The figure presents water levels converted to discharges. Since some ranges of probable water levels were presented here as mean values, the precision of the early values in particular is lower than it might appear. The interruption of large floods since the redirections of the Rivers Kander and Aare in the headwaters is obvious

Lake Constance above the mouth of the Aare also shows a noticeable gap of high levels.

- **River Main**

Thanks to the extensive preliminary studies on historical climate and change in the catchment area of the River Main by Glaser (1991 and 2008) and extensive source analyses (e.g. Seberich 1958; Glaser 1998) as well as geomorphological studies of the valley floor (Kessler 1962), very detailed information is available for the flood events of the Main and its tributaries. The preserved heritage is well illustrated by the numerous flood marks that have been preserved in the old town of Wertheim (Wehnert and Paczkowski 1985), which were compiled near the Kittsteintor on the bank of the River Tauber, which enters into the River Main here (Fig. 3.6).

In Wertheim, flood marks on buildings have a tradition, one of the oldest dating from 1595 (Fig. 3.7). At the beginning of March 1595, a flood sequence of several flood waves occurred in quick succession within a few days,

Fig. 3.6 Compilation of the flood levels of the Rivers Main and Tauber in Wertheim. (The highest marks are just above 8.50 m: 1784–1682–1732–1845–1764–1723–1595–1920–1909–1799–1809–1633–1661–1744—left 1995–1831–1658–left 1970–1820–1841–1848–1988–1830–left 1982–1844–1846)

Fig. 3.7 Historical flood marks in the historic centre of Wertheim

caused independently of each other by melting snow with ice-jams and subsequent continuous rain, leaving behind not only extensive damage and destruction but also deep scouring and expanded sediment covers in the floodplain (Glaser 2008). Even today, the historic marks are preserved during renovations and the record of the water level of the floods that continue to frequently inundate Wertheim's historic centre is updated (Fig. 3.8).

The flood level of the River Main in Würzburg can be reconstructed over a comparatively long period of time. For example, the following vivid description of the flood of 1551 has been handed down (Umweltamt Würzburg 2007, translated from German language):

> … a very heavy rain with thunder and lightning has fallen in Wirtzburg and the surrounding area, so that the Mayn has grown continuously, and small streams have opened up as if they were great torrents, causing great damage to fields and meadows, vineyards, buildings and others.

Seberich (1958) has already evaluated historic source texts on the flood records and offers an illustrative example of a source-critical analysis with his discussions and explanations of the data in footnotes. Since his estimates of the discharge are based on today's dredged Main bed and do not take into account the changes in the topography of the river

Fig. 3.8 Updated flood marks of recent events in the historic centre of Wertheim. A comparison with the flood marks collected from the entire city area (Fig. 3.6) puts the highest water level marked here from 1995 and the height of the elevated electricity distribution box on the wall into perspective

and the floodplain since historical times, only the historical water levels are illustrated in Fig. 3.9. For the period from 1845 onwards, the sporadic historical flood levels are supplemented by the first systematic records (Wallner 1957).

With regard to the human impact on flood dynamics since historic times, Wallner (1957) demonstrates that the clearing of hedges and bushes in the floodplain and along the banks of the River Main had a greater significance than the installation of groynes and guide structures. Thus, the assumption emerges that the high water levels in the Middle Ages and early modern times had their cause less in stronger floods than in the slower runoff in the floodplain, which may have led to higher water levels than today due to backwater effects (Schmidt 2002). Gerlach (1990) also found that clearing and changed cropping patterns have increased soil erosion and this has significantly influenced the sediment balance of the river. The establishment of mill weirs and the systematic river development did not override the largely natural dynamics of the river until the nineteenth century. Accordingly, evaluations of changes in flood frequency and their magnitudes since historical times must also be analysed, which is summarised below in a supraregional review.

- **River Elbe**

The floods of the River Elbe and its tributaries have received special attention in Germany, especially since the extreme flood in August 2002. The widespread flooding in Dresden, among other places, the numerous breaches of dikes and the further damage with a total volume of more than 10 billion Euros (IKSE 2004) have given renewed importance to the question of the frequency and possible magnitude of floods. However, it is precisely this event that illustrates the careful differentiation of flood parameters and reference periods for classification in the historical series, going back to AD 590 (Pötzsch 1784). As Fig. 3.10 illustrates, the 2002-flood in Dresden actually had the highest water level of both the historic and recent record.

The cause was intensive precipitation lasting several days from a so-called V_b-weather situation in which warm, humid air masses from the Mediterranean region are led northwards and glide onto colder air masses. In combination with the ascent over mountain ranges such as the Sudeten and Ore Mountains, this resulted in precipitation totals of locally 312 l per square metre and day, the highest value ever measured in Germany (BfG 2002).

Fig. 3.9 Flood levels of the River Main in Würzburg 1275–1844 (data from Seberich 1958 and Wallner 1957)

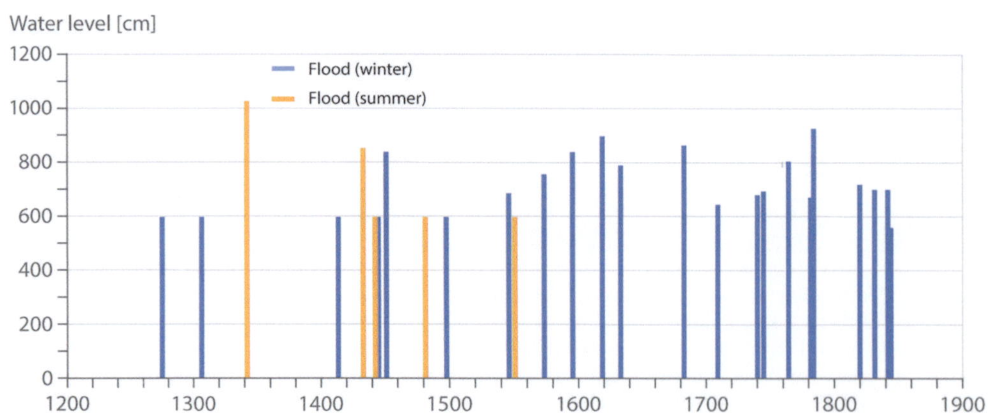

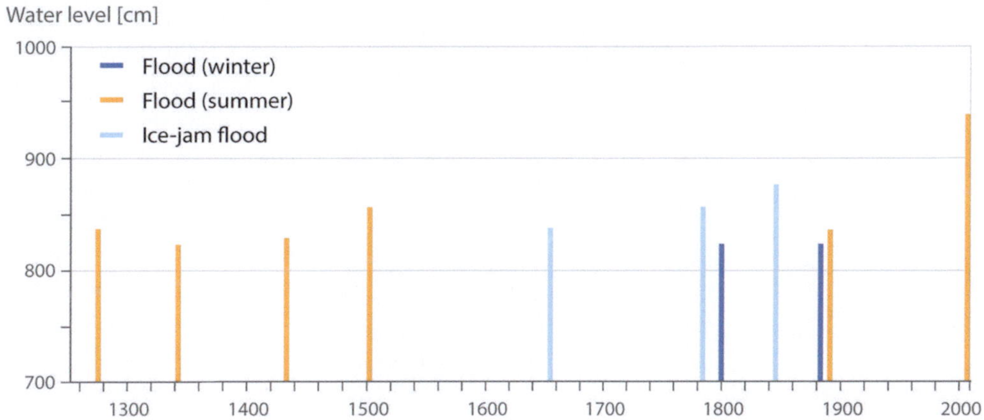

Fig. 3.10 Extreme flood levels of the River Elbe in Dresden (data from BfG 2002, not all flood events are listed)

Table 3.2 Peak discharges of the River Elbe in Dresden

	DKKV (2003) (m^3/s)	Bartl et al. (2008) (m^3/s)
1784—Ice-jam flood		4500
1845—Ice-jam flood	5700	4800
1890—Summer flood	4400	4000
2002—Summer flood	4680	

However, if one compares the discharges, the peak discharge of 4680 m^3/s for the August flood of 2002 (DKKV 2003) does not show a disproportionately high value that would justify a characterisation as a "millennium flood" (Table 3.2). Rather, a return interval in the range of 120–200 years was estimated (IKSE 2004).

But how can we explain the supposed contradiction between an unprecedentedly high flood level and a previously exceeded peak discharge? The analysis of the events and local conditions revealed that the floodplains have been strongly altered in their permeability for floods by settlements and vegetation, so that the flood discharge capacity has been noticeably reduced (DKKV 2003). This is all the more remarkable as, as consequence of the extreme floods of 1845 and 1890, flood channels were deliberately constructed in Dresden in order to be able to discharge extreme floods through the city area with as little damage as possible. In the meantime, these channels have become overgrown with bushes as part of nature conservation measures in the urban environment and their drainage capacity has been additionally reduced by deposited flood sediments. Buildings on the outskirts are also said to have reduced the capacity of the pathways, resulting in a disproportionately high risen water level. In addition, the artificially narrowed and thus underfit channel bed of the River Weißeritz, a tributary of the Elbe, was no longer sufficient, as bridges and culverts were quickly blocked by entrained flotsam and sediments and the flood moved through the adjacent streets parallel to the river channel. Consequently, Pohl (2004a, b) proposes the aimed development of traffic routes as additional flood pathways.

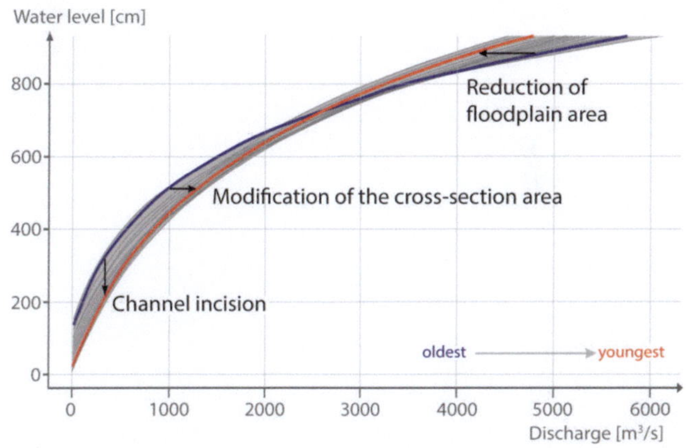

Fig. 3.11 Changes in the water level—discharge curves of the River Elbe level in Dresden due to human impact (modified after Bartl et al. 2008)

The changes in the floodplain in and since historical times are taken into account by Bartl et al. (2008) by developing time-adjusted water level-discharge curves with the aim of being able to convert the historical water levels into discharge values more easily and thus to extend the time series for the statistical analysis of the return intervals. Calibrated water level-discharge curves directly determine the discharge from the water level, which is relatively easy to measure. The change in the curves for the Elbe gauge in Dresden (Fig. 3.11) can be explained by local lowering of the channel bed, changes in the cross-section area, and the reduction of the floodplain due to embankment and building development.

In the upper reaches of the Elbe and its tributaries located in the Czech Republic, the 2002-flood had a similar devastating effect as in the German part downstream. While the water level of August 2002 at the Elbe gauge in Děčin had been exceeded by previous floods similar to Dresden (IKSE 2004), the tributary Vltava River in Prague showed a peak level never reached before (Fig. 3.12).

Within the framework of the reconstructions of peak discharges of historical floods in Prague, Elleder et al. (2013)

Fig. 3.12 Gauge of the Vltava River at Charles Bridge in Prague. The flood level of August 2002 has not been reached by the previous ice-jam floods of February 1784 or March 1845, nor previously by the maximum summer flood level of September 1890. The face of Bradáč, the bearded man, is used as a relative reference point of the water level

document that, unlike in Dresden, also no larger discharge has been observed before in historical times (Fig. 3.13). With a measured peak discharge of 5160 m^3/s, the August flood 2002 of the Vltava in Prague also exceeds the ice-jam floods of 1784 and 1845 with their estimated peak discharges of about 4500 m^3/s each (Brázdil 1998).

3.1.2 Historic Floods in China

Due to the long history, China has some of the oldest descriptions of floods. Pang (1987), for example, shows that reports date back to around 2000 BC, even if, as expected, they describe more the damage that occurred and the actions of the respective rulers than the characteristics of the flood itself. Since systematic gauge records were only made after the founding of the People's Republic of China, one has to rely on the evaluation of historical data for long-term analyses. Individual flood marks go back as far as AD 223, although more frequent records begin around 1000 years later (Fucheng 1987; Luo 1987). Floods in China received worldwide attention in the twentieth century. Large areas of eastern China, namely the fertile lowlands along the Huang He, Huai and Yangtzekiang Rivers, were repeatedly regions of catastrophic floods (Jiaqi et al. 1998). In 1931, probably the most fatal flood in the world occurred here, when large regions were inundated for several months and only individual centres rose out of the water as islands, ultimately costing the lives of an estimated 145,000–4,000,000 people. But this entire region is by nature strongly characterised by floods, as a glance at the two main Rivers Huang He and Yangtzekiang illustrates.

The River Huang He, also called the Yellow River, rises on the eastern edge of the Kunlun Shan Mountains, flows through the loess hill country and along the south-eastern margin of the Gobi Desert, and finally flows into the Yellow Sea. The river's name is derived from its high sediment load, which is due to strong erosion in the loess hill country of the river (Brush et al. 1989). This is also the core problem of the recurrent floods in the lower reaches of the Yellow River, because the decreasing gradient in the lower reaches and the reduced drag force cause the sediment load carried to be deposited. Over time, this causes the river channel to rise above the surrounding land surface (Fig. 3.14).

The rise of the Yellow River bed by more than 10 m in a century, illustrated in Fig. 3.14, is accompanied by a gradual formation of natural levees. These form from the entrained suspension load deposited immediately adjacent to the river channel during overbank flow (Klasz et al. 2014). The Huang He flows 3.5 m above the adjacent land surface (Changming and Jiyang 1989). The dramatic effects of flood discharges become obvious due to the high elevation of the river course. Large areas are suddenly inundated as soon as the water level exceeds the natural levees and cannot drain back into the river channel after the flood wave has passed through. The "Great Plain", as the extensive

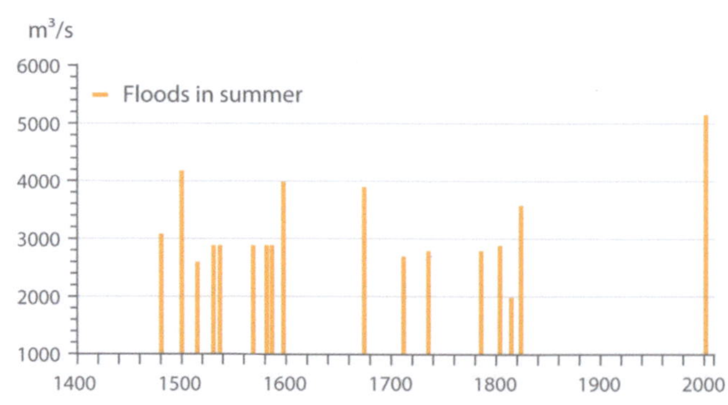

Fig. 3.13 Reconstructed peak discharges of extreme summer floods of the River Vltava in Prague (modified after Elleder et al. 2013)

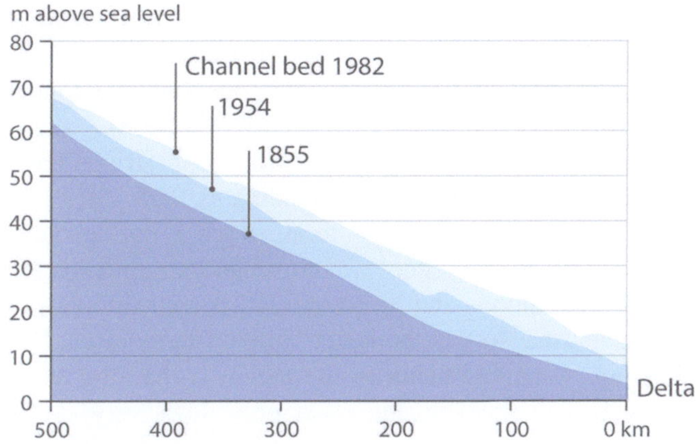

Fig. 3.14 Longitudinal profile of the lower reaches of the River Huang He (modified after Qingchao 1989)

lowlands in eastern China between the lower reaches of the Huang He and Yangtzekiang Rivers are called, is characterised by continuous large-scale displacements of the rivers and massive accumulations of deposited flood sediments (Fuling 1989; Xue 1993; Saito et al. 2001; Zhang 2009). Here humans also intervene through their influence on soil erosion in the middle reaches, which has led to a continuous increase in sedimentation in the lower reaches over the last 2300 years (Xu 2003). Among the thick deposits are archaeological traces from all periods of Chinese history, with excavations at Sanyangzhuang receiving particular attention (Lawler 2010). A 70 cm thick layer of clay was deposited over a wide area here around 2000 years ago, which means that all the details of everyday rural life in this period have been preserved, earning the site the name "Pompeii of China".

A sad status as a dramatic environmental war measure has been achieved by the National Chinese troops' deliberate opening of the embankments and dams on the Huang He during the Sino-Japanese War in 1938, which is hardly known in the West (Dutch 2009). In order to stop the advance of the Japanese troops, the extensive lowlands were deliberately flooded. Militarily, this drastic measure actually brought a slight advantage; on the other hand, the flooding itself as well as the subsequent epidemics and famine killed over 800,000 people. But this was also not the first case of deliberate opening of the dikes on the Yellow River. In 1128 and 1642, the inundation of the lowlands was used as a weapon at Kaifeng. In the first operation, an attacking army was also repelled, which was successful, but involved a shift in the course of the river. This resulted in great territorial losses, as the course of the river marked the border. In 1642, the dikes were opened during a peasant uprising, although it is not clear whether this was to keep the revolting peasants away from the city or whether the attackers wanted to overcome the city's fortifications.

The situation on the Yangtze River, which is also known as the Chankiang River, is similar to that on the Yellow River. Here, too, a steadily increasing flood problem is observed, which can be traced back to natural aggradation of the riverbed combined with the formation of natural levees (Yin and Li 2001; Jiang et al. 2005). The flood retention effect of some lakes in the upper and middle reaches of the river is also reduced by their silting up, which is planned to be compensated by the famous Three Gorges Dam (Fu et al. 2010). The valley floor is characterised by thick deposits that contain traces and remains of numerous cultural stages but also of striking flood events (Fig. 3.15).

The sequence of layers documents repeated flood deposits carried in by the Yangtze River, which could be dated by adjacent archaeological finds or dating of charcoal deposits using the radiocarbon method. In this way, evaluations of historical descriptions (Jiaqi et al. 1998; Zhang et al. 2008), many of which were unfortunately lost during the Cultural Revolution in China, can be supported.

The chronology of flood and drought periods in China based on historical records shows an often contrary picture for the Huang He and Yangtzekiang (Fig. 3.16).

Frequently, the rivers experienced contradictory pattern of periods dominated by floods respectively droughts. In addition, indications of periodic changes with frequencies of 100 and 400 years are seen in the indexed hydrographs. One explanation for this can be seen in changes in the monsoon, which is decisive for the climate and the water flow of the rivers. Its abrupt changes are now known (e.g. Morrill et al. 2003). This may also explain why the Huang He had peak discharges of 43,000 m^3/s in the early Holocene, while the peak discharge in historical times (AD 1843) was only 32,000 m^3/s, which anyhow obviously exceeds the maximum value of the measurement period of 5637 m^3/s (1994) (Yang et al. 2000).

3.1.3 USA

The USA represents a counter-extreme to the millennia-long historical record of floods in China. There are no written records from the time before the settlement of the European immigrants, and since this period is relatively short, especially in the western parts of the country, there are practically no historical flood records (Hoyt and Langbein 1955). Against this background, indirect methods for flood estimation are being further developed, because flood events regularly exceeded the short gauge records, which challenges flood magnitude estimations (Costa 1987a, b). Added to this are influences from cyclones in the southeast of the country, flash floods in the semi-arid and continental regions, as well as atmospheric circulation fluctuations such as in the periodically occurring El Niño and La Niña years, in

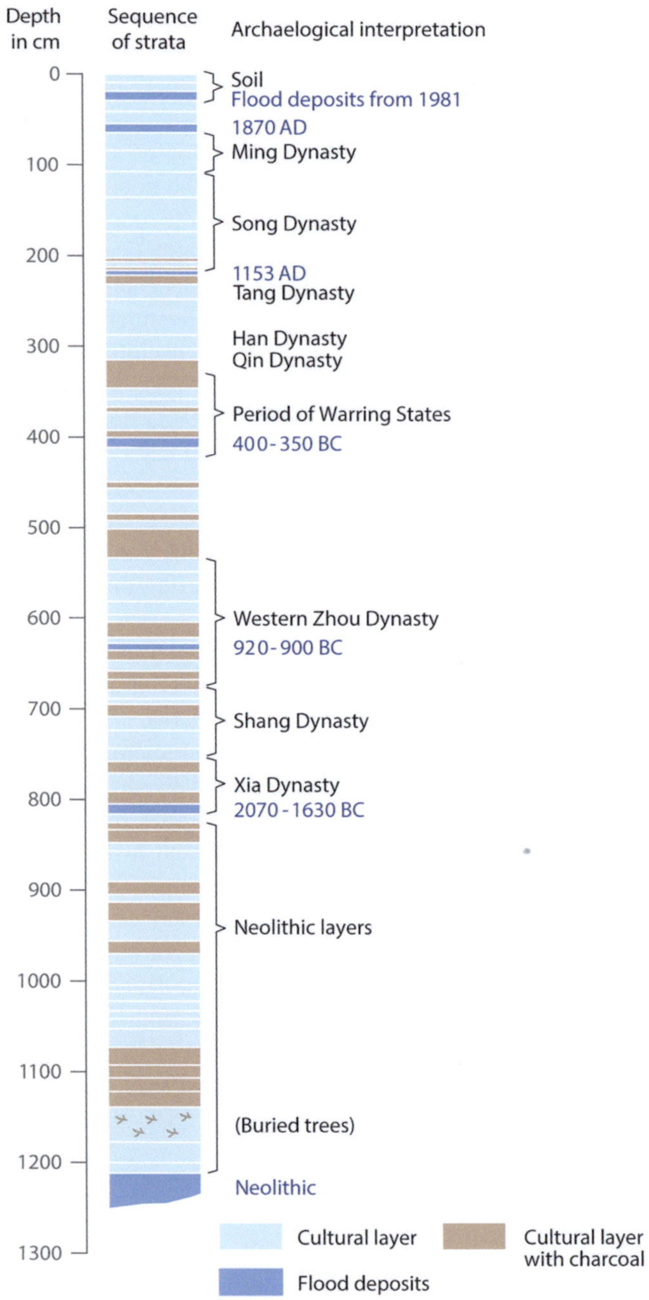

Fig. 3.15 Profile of flood deposits and archaeological layers at Zhongba, Three Gorges Reservoir (simplified after Cheng et al. 2005)

which above-average precipitation falls e.g. in California. Such boundary conditions lead to extraordinary flood events, which are not always easy to cope with and control (O'Connor and Costa 2003; Costa and Jarrett 2008).

In April and May 2011, the Mississippi River basin experienced repeated extreme flooding (USGS 2013). At the time of the annual spring flood, which is due to snowmelt, several storm systems accompanied by numerous tornadoes, flash floods and heavy rainfall passed through the catchment area, so that the water volumes to be discharged added up. A critical factor in the region is that due to the extremely low gradient, the water only flows very slowly into the Gulf of Mexico. Another complicating factor is that the geometry of the catchment area, which drains more than 1/3 of the contiguous US states, is shaped in such a way that flood waves from the large tributaries overlap in the lower reaches. In May 2011, for example, the flood discharge at Vicksburg reached 66,000 m^3/s, whereas 16,000 m^3/s is typical. Due to the unfavourable natural conditions, the flood was not unusual. The floods of 1927 and 1993, for example, had comparable dimensions, except that people were not adequately prepared beforehand, so that very great damage was caused (Galloway 2004). The 1927 event in particular left its mark on the country, as the damage amounted to about 1/3 of the US budget, making it the most destructive flood in the USA in historical times (USGS 1995; Barry 1997). On the other hand, this marked the beginning of systematic safety measures along the river courses, so that subsequent floods could be better controlled. Whereas in 1927 uncontrolled breaches of dams were still the rule, in 2011 targeted river diversions over agriculturally used areas were able to come into operation, thus avoiding damage in the major cities, namely also in New Orleans, which was devastated by Hurricane Katrina in 2005. On the other hand, recent studies considering historic floods in addition to gauge data indicate, that Mississippi River flood hazard is amplified by river engineering (Munoz et al. 2018).

The question arises as to whether these events already represent the maximum flood magnitudes. As mentioned above, there is a gap in our knowledge about flood events

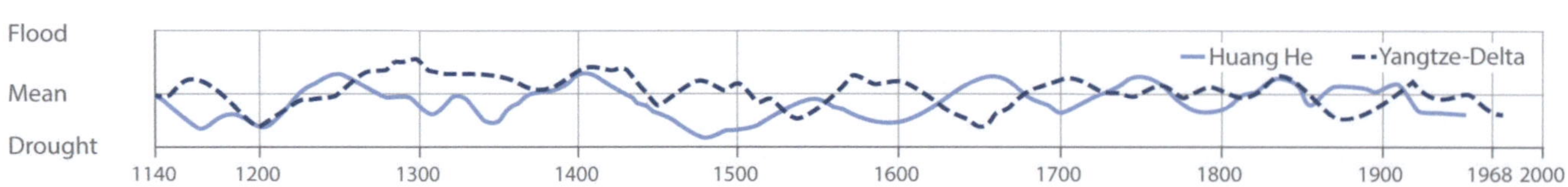

Fig. 3.16 Comparison of the occurrence of floods and droughts on the Huang He and Yangtzekiang (modified after Jiaqi et al. 1998)

before 1927 due to the lack of quantifiable historical records (Smith and Reed 1990), which is not significantly changed by individual impressive flood descriptions from remarkably early times. Garcilaso de la Vega, for example, was probably the first white man to see the Mississippi in 1543. He handed down an impressive description of an oncoming flood and the extensive flooding that lasted 40 days on the wide flat valley floor (after Frankenfeld from Barry (1997, p. 173)):

> Then God, our lord, hindered the work with a mighty flood of the great river, which... came down with an enormous increase of water, which in the beginning overflowed the wide level ground between the river and the cliffs, then little by little it rose to the top of the cliffs. Soon it began to flow over the fields in an immense flood, and as the land was level, without any hills, there was nothing to stop the inundation. On the 18th of March, 1543,... the river entered with ferocity through the gates of the town of Aminoya. It was a beautiful thing to look upon the sea where there had been fields, for on each side of the river the water extended over twenty leagues of land, and all of this area... nothing was seen but the tops of the tallest trees.... These floods occur every fourteen years, according to what an old Indian woman told us, which can be verified if the country is conquered, as I hope it will be.

Obviously, the repeated widespread flooding—20 leagues correspond to about 90 km—was not an obstacle for the Indian population to establish permanent settlements either, as this description from the region near Greenville not far from the mouth of the River Arkansas illustrates. As a result of repeated widespread flooding, thick alluvial clay layers were deposited in the wide lowlands of the lower Mississippi, similar to the catchment areas of China's major rivers, which contain evidence of flood dynamics even in prehistoric times. On the other hand, the flood intensity of the lower Mississippi River, which fluctuated over longer periods of time, had consequences for the early settlement of the region. Thus, changes in channel bed patterns and valley stage formation of the Mississippi and Arkansas Rivers correlate with breaks in settlement and economic patterns of early populations observed in the periods AD 1000–1200, 1000–500 BC, and 2800–1800 BC (Kidder et al. 2008). This fits temporally with observations of shifting agricultural settlements in the Grand Canyon on the Colorado River (Anderson and Neff 2011). Here, too, striking changes are evident in the period AD 1050–1170, when the settlement sites were successively elevated, moving away from the course of the river. In this context, it is questionable whether the first Native American settlers first had to gain experience with the flood levels of the Colorado or whether its flood magnitudes increased significantly precisely during this period, as the compilations of increased flood activity in the southwest of the USA in the Holocene indicate (Ely 1997; Harden et al. 2010).

A complicating factor in such studies is that large parts of the West and especially the Southwest of the USA have semi-arid to arid semi-desert and desert climates and therefore are characterized by very high variability of hydrological conditions (Ingram and Malamud-Roam 2012). This is characterised by only episodic runoff in the river courses, which is fed by flash floods in addition to snowmelt from higher elevations in spring. These flash floods are formed by convective precipitation typical of thunderstorm cells, from which large amounts of precipitation can be concentrated in a short time over a small area. In contrast to "normal" flood events, these flash floods show a much higher destructive energy content with shooting runoff, even if their peak discharge is lower (Costa 1987a, b; Negm 2020). The resulting disproportionately high shear stress is capable of displacing large amounts of sediment in a very short time, causing the streams and rivers to change their appearance fundamentally and abruptly, making it difficult to reconstruct the discharge from the sediments. Prominent examples of this are the controversial discussions about the discharge in Bronco Creek, north-western Arizona, on 19th August 1971 (House and Pearthree 1995; Costa and Jarrett 2008). After a thunderstorm rainfall, the runoff there increased drastically and showed strongly pulsating waves that flooded a road bridge in 4–5 min intervals over a period of 2 h. First reconstructions resulted in record discharges, which classified this flood as the world's strongest event ever measured in terms of specific discharge of over 55,000 l/(s km^2). As a comparison with maximum discharge values in Germany illustrates, this value is higher by a factor of 10 than the values observed in Germany for catchments of similar size of 30 km^2 (cf. Chap. 2). The comparison is not actually permissible in this way because of the necessary regionalisation, but it illustrates quite clearly the magnitude of the event in the desert area. The attention that this event received as a result of the classification led to follow-up investigations to verify the estimate. This revealed problems in determining the flood level and discharge because, on the one hand, the crests of the periodically passing waves of the flood are not representative and, on the other hand, the cross-section area cannot be clearly determined due to the sediments carried along and aggradation of the river channel. Questions arose as to whether such an event a peak discharge with a further meaning at all or whether one could only speak of short-term periodically fluctuating water levels and how this phenomenon of waves should be assessed at all (Hjalmarson and Phillips 1997; House et al. 1998). Summarising the discussions, Costa and Jarrett (2008) suggest that, because of the exceptional discharge phenomenon, the maximum discharge when one of the waves passes should be considered as the peak discharge, but that this value should not be taken into account when drawing up comparisons or carrying out regionalisation. While no direct measurements are available for the Bronco Creek flood, gauge data from Tanque Verde Creek, located on the eastern outskirts of Tucson, Arizona, show how abruptly the discharge can rise during flash floods (Fig. 3.17).

At the end of July 1981, several thunderstorm cells passed through southern Arizona, none of which brought above-average rainfall (Hjalmarson 1984). Although a flash flood had already passed through in the early morning hours of the previous day, in the late afternoon of 26th July numerous people were again bathing in the pools at Tanque Verde Creek (Fig. 3.18), a popular excursion and hiking destination in the desert. Eyewitnesses reported that within 15 s the water level rose by half a metre, and in less than a minute the water was over a metre deep. The front of the flash flood with the stones and logs it carried came faster than the bathers could flee. In the process, 8 people were killed.

It becomes obvious that variability is an essential parameter in flood estimation in the USA. In addition to episodic precipitation in desert areas and periodic changes in atmospheric circulation with drastic changes in precipitation amounts in the context of El Niño and La Niña years, evidence has recently emerged that there may also be even longer-term circulation fluctuations that lead to extreme floods. In January 1862, for example, numerous rivers in California experienced floods that were accompanied by the highest peak discharges and water levels on record (Sidler 1968; Clarke 1996; Engstrom 1996). This was preceded by weeks of precipitation resulting from an unusually steady succession of low-pressure systems passing through, which meant that sunlight breaking through the cloud cover for a few minutes became a newspaper headline in southern California. Although little quantifiable information has survived due to the sparse settlement at the time, it is certain that no such precipitation in terms of quantity or duration has ever been observed and reported. As fishermen reported, a freshwater layer about 1 m thick formed in San Francisco Bay from runoff precipitation water on top of the heavier salty seawater, the outflow of which through the Golden Gate into the Pacific superimposed the otherwise cyclical changing tidal current. Furthermore, virtually the entire Los Angeles Basin was flooded with a depth of about one metre. The thickness of the flood deposits was almost sufficient to cover locally planted vines, which subsequently meant that only irrigated agriculture could be practised in the permeable sandy cover. The subsequent extensive development of the region has largely covered the traces of the flood, but a peak discharge of around 9000 m^3/s has been estimated from records of local water levels and the channel bed geometry of the Santa Ana, which drains large parts of the Los Angeles area (Sidler 1968). This value is three times the highest measured discharge from 1932. This event, described as the "biblical deluge among the floods of California", can definitely not be correlated with an El Niño year, although the assumption of a connection with the end of the Little Ice Age is also only speculative (Engstrom 1996). If such an event

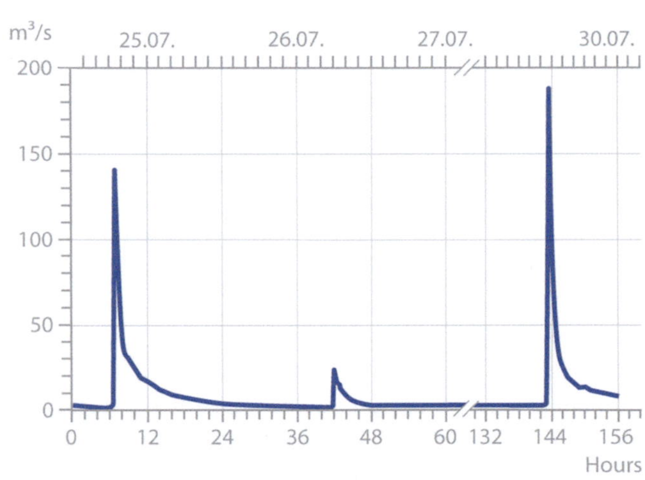

Fig. 3.17 Discharge of three flash floods in Tanque Verde, Tucson AZ, July 1981 (data from Hjalmarson 1984)

Fig. 3.18 Valley of Tanque Verde Creek, east of Tucson, Arizona, USA

were to occur today, the sealing of the surfaces by buildings in the meantime would result in changed, probably more extreme runoff dynamics. Expected damages and economic consequences have been developed as detailed estimates (Porter et al. 2011). These estimates are based on the assumption that special trajectories of increased moisture transport ("atmospheric rivers") develop in the atmosphere, which can lead to regionally above-average precipitation, as observed in California in the winter of 1861/62.

Nilometer—The Oldest Systematic Flood Records of Mankind

The oldest systematic flood records were carried in Ancient Egypt by the gauging station of the so-called Nilometer (Popper 1951; Said 1993). Since 3018 BC, the flood level of the River Nile was recorded to calculate taxes from the local farmers. The annual floods were directed into the arable land to wash out the salt, irrigate the plants and distribute the fertile loam. First river channel modifications and deviations including flood protection measures around the historic city of Memphis were carried out. A flood record is handed down from Cairo for the period AD 622–1921, which is unfortunately more complicated to interpret than it might be assumed at a first glance. Like several other large river systems, the Nile is characterized by aggradation in the lower parts due to deposition of the transported sediments (Sutcliff and Parks 1999; Woodward et al. 2007). Considering the timescale of several centuries, the increase of mean and maximum flood levels has to be put in perspective. Unfortunately, discharge estimations based on the recorded water levels like for other historic flood marks are challenged by the hard to quantify cross-section area due to the numerous irrigation canals that have constantly been modified.

3.2 Remarkable Flood Events in Central Europe

Despite the abundance of floods in historical times, individual events are still particularly remembered centuries later. Be it because they reached particularly extreme water levels, had a supraregional impact or caused lasting damage.

3.2.1 Magdalen Flood 1342—The Ultimate Flood Disaster in Central Europe

Along numerous rivers in Central Europe, exceptionally strong floods occurred in July 1342, which are therefore

Fig. 3.19 Compilation of flood levels of the River Weser at Hannoversch Münden

called Magdalen floods due to their onset around the 22nd in accordance with the calendar used at that time after the church holidays. In many places the most extreme flood levels in historical times were reached, for example in Würzburg (Fig. 3.9), Frankfurt and Hannoversch Münden (Fig. 3.19), which is why the event has repeatedly been characterised as the millennium flood (Tetzlaff et al. 2002; Dotterweich and Bork 2007; Zbinden 2011; Bauch 2014).

Numerous text sources dealing with water levels, damage and the spread of the flood have been handed down (e.g. Weikinn 1958; Alexandre 1987; www.tambora.org). The following contemporary quotation is an example of the deep impression that the flood levels made on people (Weikinn 1958, translated):

> Immediately afterwards, also in the year 1342, on 21st July on Praxedist Day, i.e. a Sunday or the day before St. Mary Magdalene's Day, from the beginning of the day and in the morning, the Main River in Würzburg rose so much above its banks that the stone and magnificent bridge in Würzburg with the towers and their walls and the city walls and also many stone houses there and all around suddenly collapsed. The flooding of the Main was such that the water flowed over the steps of the portico of the church in Würzburg near the stone statues to an unusual extent. All the wooden and stone bridges above and below the Main also collapsed. Also, tremendous

> damage occurred to the city and farmer's estates along the Main everywhere, deplorably.
>
> Indeed, in other parts of the world, too, there were unheard-of floods of the waters. Also, all underground springs of water were broken, as it were, and the floodgates of heaven were open, and rain fell upon the earth as in the 600th year of Noah's life, as you can read about the Flood in the 7th chapter of Genesis in the middle.

For the area of the Kleinheubach gauging station, located in the lower Main valley near Miltenberg, Tetzlaff et al. (2002) estimate a discharge of 3500 m^3/s, which corresponds to about twenty times the mean discharge of 172 m^3/s and is associated by the authors with a return interval of 10,000 years. Further downstream at Frankfurt, detailed descriptions of the water level in various churches have survived (according to Grotefend in Bork and Kranz 2008), which make it possible to determine the water level there at 7.85 m above the present-day gauge zero. The situation is similar in Hannoversch Münden, located at the confluence of the Rivers Fulda and Werra and called River Weser from thereon downstream. Based on surviving water level descriptions, the water level in July 1342 was around 8 m above the mean water level, thus inundating the entire historic centre of the town (Fig. 3.19).

The cause of the flood was a precipitation event that has not yet occurred with such intensity in Central Europe. The weather conditions of the year can be reconstructed from various sources (e.g. Tetzlaff et al. 2002; Bork et al. 2006; Glaser 2008) and for sure, there was more than one flood event in the year 1342 in Europe, which occasionally leads to confusion for the reconstruction. According to these, the first flood occurred at the end of a partly severe, but generally average winter after snowmelt at the end of February. This, too, had widespread distribution in various catchment areas and caused damage in large parts of Central and Western Europe and destroyed bridges in Dresden and Prague (Brazdil et al. 2005). However, the defining event in the summer of that year resulted from a weather situation that is often compared to a V_b weather situation, as was responsible for the floods of the Oder in 1997 and the Elbe in 2002. The spatial arrangement of high-pressure and low-pressure areas over Europe causes warm and humid air from the Mediterranean region to be led northwards, where it meets cold air masses and forces them to rise, resulting in heavy precipitation. The centre of the precipitation seems to have been over the Main valley, as most reports are available from here and the adjacent upper catchment areas of the neighbouring rivers. Estimates show that more than half of the current annual precipitation fell here within about one week (Glaser 2008).

The consequences of the heavy precipitation were catastrophic, as the subsoil could not absorb the water levels, which led to severe erosion. Evidence of up to 10 m deep gully erosion and intensive soil displacement attributed to the precipitation in July 1342 is available from the Lower Eichsfeld region south of the Harz Mountains (Bork et al. 2006) (Fig. 3.20), the Spessart region (Bork and Kranz 2008) and northern Bavaria (Dotterweich et al. 2003), among others. Bork et al. (1998) estimate the soil losses

Fig. 3.20 Remaining gully erosion related to the Magdalene Flood in July 1342. The age of the gully is derived by ceramics within the re-fill of the valley located near the village of Obernfeld (20 km east of Göttingen) dated as fourteenth to fifteenth century AD (Bork et al. 1998, 53f)

associated with the heavy rainfall of 1342 to be 13,000 × 10^6 t, which corresponds to a mean soil loss of 23 mm, a value many times higher than the mean annual values under present-day conditions at the time of large-scale and erosion-prone intensive agriculture (50 × 10^6 t or 0.08 mm per year). If we take the traditional chronology of precipitation and floods as a basis, the precipitation area of Central and Upper Franconia (19/20th July) has slowly shifted to the north-east, as the reports of heavy precipitation in Thuringia and Saxony begin one day later. According to the orientation of the river courses, the flood waves appear with a time delay: Frankfurt on 20th, Mainz on 21st and in Cologne on 25th of July. According to the later onset of precipitation, the flood wave reaches the lower Werra and the Weser valley on 22nd July, while in Frankfurt the flood lasts until 24th of July. The distribution map shown in Fig. 3.21 is developed from the source texts on extreme precipitation and damage descriptions of floods.

The distribution of the extreme precipitation and the resulting flood shown on the map does not correspond to the large-scale map representation as advocated elsewhere

Fig. 3.21 Area of the Magdalene Flood in July 1342 (Herget et al. 2015). In the coloured region around the headwaters of the Rivers Main and Weser, there was extremely abundant precipitation, which led to exceptional flood levels in the named locations. Note the locations with high magnitude flooding, but without significant local precipitation

(Brazdil et al. 2005, Bork et al. 2006). The background to this is that some source texts require critical examination and cross-referencing before they can be adopted for the reconstruction of the events in July 1342. This will be exemplified by source texts in Weikinn (1958) and Alexandre (1987), even though a complete analysis and discussion is not possible here. If one focuses on source texts in which floods and flood damage are explicitly assigned to the event in July 1342 and not to the year 1342 in general, the distribution area results in a nevertheless coherent overall picture. For example, the flood damage in Prague and France in 1342 is associated with the February flood of 1342, while the flood damage in Lombardy dates to November 1342. The possibility of crossing the city walls of Cologne by boat during the flood is often cited, although this can demonstrably be attributed to the flood of 1374 (cf. following chapter below). On the other hand, however, a source text is passed over that explicitly points out that the flood in the River Rhine resulted from the correspondingly increased inflow of the River Main or that the flood in Limburg an der Lahn occurred without increased precipitation. This illustrates indications that the precipitation area was spatially limited and that flood mentions of the River Eider in far north of Germany in Schleswig–Holstein need not be causally linked; on the contrary, a storm surge of the North Sea has also been handed down for this year (www.tambora.org). This is also the direction taken by Brazdil et al. (2005), who point to the absence of any mention of increased precipitation or flooding in immediately neighbouring Bohemia. The same applies to the Ore Mountains and Saxony, the Carpathian Basin as well as the catchments of the Rivers Oder and Neckar, of which no mention is made in regional overviews of July 1342, although this would be expected (Röckel 1994; Fügner 1995; Schmidt 2000; Pohl 2004a, b; Kiss 2009, 2019). Furthermore, it is also explicitly noted for the Eastern Alps or the area of Austria that this period does not appear in particular (Rohr 2007) or is not clearly assigned to the summer of the year (Röthlisberger 1991). Findings from Switzerland hitherto assigned to the Magdalen flood, such as the depictions from Lucerne, according to which the high altar of the Barfüsserkirche was reached by the flood, date from 29th June 1343 (according to Schaller-Donauer from Siegenthaler and Sturm 1991). Finally, Roth (1996) already suspected that the precipitation event in July 1342 might not have been a consequence of a low moving northwards on a V_b circulation pattern like suggested elsewhere, but a quasi-stationary front. Herget et al. (2015) demonstrate, that also convective precipitation event could have been possible.

These aspects lead to the hypothesis, still to be verified, that the extremely intensive precipitation-producing low pressure area had only a regionally limited distribution (Fig. 3.21). Possibly, the precipitation had been limited to the catchment area of the Main and the upper reaches of the adjacent catchments and subsequently shifted northwards to cause gully erosion south of the Harz Mountains (Fig. 3.20). From this region, the resulting flood waves subsequently poured out along the respective river courses. The floods on the Lower Rhine, the Rivers Weser and Danube would then have arisen from tributaries from the Rivers Main, Fulda and Werra or Altmühl, without the flood waves in the regions themselves arising from heavy precipitation.

3.2.2 Winter 1784—An Ice-Jam Flood Makes History

At the end of February and into March of 1784, floods occurred throughout Europe that are undoubtedly to be counted among the extreme events in historical times. Accounts such as the following one from Cologne (Thelen 1784, translated) (Fig. 3.22) illustrate the lasting impression left by the floods:

> The winter of the present 1784 year will forever be called the terrible one in our yearbooks, and no one who has lived it can think of it without shuddering. Ghastly devastation, extreme poverty, black hunger, violent death, fears of death more terrifying than death itself, are the cruel features that will forever distinguish it among centuries; fierce cold, and extraordinarily frequent snow prepared the tools of its fury, ice and flood."

Ausführliche Nachricht von dem erschrecklichen Eisgange, und den Ueberschwemmungen des Rheines, welche im Jahre 1784 die Stadt Köln, und die umliegenden Gegenden getroffen.

Mit Obrigkeitlicher Genehmhaltung und Einsicht herausgegeben von Johann Leonard Thelen, Professor des hiesigen Lorenzianer Kollegiums.

Mit Stadtkölnischem Privilegio exclusivo.

Erstes Stück.

Fig. 3.22 Cover of the contemporary descriptions of the ice-jam flood of 1784 in the Cologne area by J.L. Thelen

Since the flood only occurred a good 250 years ago, much more detailed accounts of the event and the resulting damage are available compared to earlier events. In addition, systematic weather observations had already been carried out and there was a Europe-wide organisation, the Societas Meteoroligica Palatina, which operated a synoptic meteorological measuring network (Kington 1974, 1988). Brázdil et al. (2010) and Munzar et al. (2005), for example, offer summaries of supra-regional overviews; further contemporary representations similar to those from the Cologne area by Thelen (1784) have, been compiled, evaluated and verified e.g. by Glaser and Hagedorn (1990) and Glaser (2008) for the River Main valley, Fricke (1988) for Heidelberg and Heuser-Hildebrandt (2005) for Mainz. The following overall picture of a Europe-wide flood event emerges. However, now that the supraregional spatial distribution of simultaneous floods has been critically questioned in connection with the 1342 flood, how can one be sure of the Europe-wide spread of the event? The answer to this question lies in the cause of the flood. For example, the event of late winter 1784 was an ice-jam flood, in which snowmelt and thawing ice sheets on river courses play a central role. Characteristic of this is the influence of ice floes that originate from a thawing ice cover on a river. These initially drift downstream with the current, but can accumulate at narrow points, river bends and, in particular, bridges and mill weirs, and are set up by the water pressure. This causes a backwater of the river, which is accompanied by a disproportionately high water level (Fig. 3.23).

At the same time, the discharge downstream of the ice jam decreases due to the backwater of the water above. However, this only happens until the floes yield to the pressure of the backwater and releases it by as a prominent wave. Beltaos (2008) suggests the term "jave" for "jam released wave" for this pulse. Due to their mass and inertia, the floes floating on the surface have a particularly destructive effect on all structures on the banks and in the river and, due to undercutting of the banks, also led to landslides during the ice flood of 1784, for example in the River Main valley near Karlstadt (Rutte and Wilczewski 1983, p. 122).

The year 1784 still belongs to the period of the so-called Little Ice Age, which lasted from the end of the Middle Ages until the late nineteenth century and was significantly colder than today (Grove 2004; Glaser 2008). But even for this cold period, the winter of 1783/84 was exceptionally severe in Europe, which is associated with the eruption of the Laki volcano in Iceland. This released large amounts of sulphur dioxide, which led, among other things, to striking observations of high-altitude smoke in Central Europe (Thordarson and Self 2003). Due to long-lasting frost, a thick snow cover accumulated in large parts of Europe, even with only average winter temperatures, practically all rivers froze over. In the meantime, atmospheric circulation

Fig. 3.23 The gauge at Cochem on the River Moselle. Note the disproportionately high mark of the water level of 28.2.1784 between the two tower windows compared to the modern water level bar and the other markers of historical flood levels

Fig. 3.24 Occurrence of ice-jam floods in February/March 1784 in Central Europe (compiled from numerous sources from Herget 2012)

fluctuations also caused temporary warm-air intrusions, which led to several supraregional floods (Brázdil et al. 2010). The most striking event of these warm air irruptions occurred at the end of February to the beginning of March across Central Europe. Locally, temperatures rose by 10 Kelvin within one day, which was also accompanied by locally precipitation of > 40 mm and caused the snow cover to thaw over a wide area. On the basis of the available weather records and numerous surviving observations, the pattern of the occurrence of ice floods on numerous river courses in Central Europe can be deduced from Fig. 3.24.

Brázdil et al. (2010) supplement the region shown in the map (Fig. 3.24) with a pan-European overview. Thus, the ice floods in northern France started on 23rd of February and began on the 27th in the Lower Rhine. Depending on local conditions with slight variations, Thuringia and Saxony were also reached by the end of the month, while the outgoing flood waves of the Rivers Elbe and Danube, for example, reached the lower reaches at Magdeburg on 4th of March, Hamburg and Vienna, respectively, by 8th of March, and Budapest by mid-March. The Alpine region was not included, apart from a few individual cases on the lower Inn and Lech (Brázdil et al. 2010; Wetter et al. 2011). This can possibly be explained by the seasonally low water level there, which may also freeze, but finds an empty and free riverbed during ice flow, in which less damage occurs. There are no relevant descriptions of damage from the Upper Rhine either—perhaps a connection can be seen here with the course of the Rhine, which was still divided into numerous arms, because the corrections to the Upper Rhine by Tulla were not carried out until later.

The damage caused by the ice-jam floods was immense; numerous bridges and mills, as well as the buildings and dykes near the river, were destroyed by the outflowing ice floes or the trees frozen in them. Tied-up ships, which had not been underway for a long time because of the ice and were frozen solid in the ice, had a similar effect. They tore themselves loose together with their moorings and destroyed many a harbour facility in the process. Elleder (2010) illustrates this using the example of the reconstructed flood hydrograph of the River Vltava on 28th of February 1784, since in Prague, for example, the water level rose by 4.5 m within 12 h at an average rate of 30 cm per hour. The rapidly rising flood waters that began locally at midnight left no time to free the ships from the ice, so that they were torn loose, carried away and destroyed in the

Fig. 3.25 Entrance gate of the church Maria Lyskirchen in Cologne, with historical flood marks of ice-jam floods of the River Rhine in 1432 and the exceptional level from 28th of February 1784

ice floe jam, as is repeatedly described of all affected river courses with navigation. The exceptionally high water level also threatened buildings in the cities where, due to their relatively high location, a flood threat had never been considered for the foundation (Fig. 3.25).

The complicated hydraulic conditions—backwater and backwater with slow flow through the ice floe barrier on the one hand and surge-like release when this barrier fails on the other hand—make it very difficult to estimate the discharge during ice floods. Accordingly, representative water levels or discharges are rarely listed in numerous flood records for ice floods. Accordingly, a more extensive estimation of return intervals must also be viewed critically, even if corresponding data are available locally (Glaser and Hagedorn 1990). Due to the anthropogenic warming of rivers in Central Europe caused by the discharge of cooling water and wastewater, long-lasting thick ice sheets no longer form on most rivers in Central Europe and no ice jam floods are observed. As illustrated by the overview figure on the cause-related differentiation of floods in the introduction of the book, ice floods in other regions of the world also have recent peak discharges that can exceed regular events by orders of magnitude.

3.3 Frequency of Floods in Historical Times in Central Europe

The chronologies and extreme events in historical time presented as examples before almost inevitably lead to the question of trends and return periods of increased respectively decreased flood frequencies. Be it in terms of frequency and magnitude in historical time, especially in regional comparison, or prognostically for the future, beyond scientists especially the insurance business is interested in these aspects to calculate the related risk. Consequently, numerous studies were carried out in recent years while by a closer look several fundamental problems are not considered, in some cases were even ignored:

- When looking at the examples listed above, it becomes clear that the problem of comparability of the flood studies carried out arises here. Often, against the background of incomplete records of all flood events, only extreme examples were listed, for example when neighbouring urban areas were flooded or serious damage occurred. But one must be aware that countless (sic!) smaller events have not been included in the chronology at all and thus potentially existing periods of numerous small floods are underrepresented to the detriment of isolated large floods (cf. Paprotny et al. 2018). Note, that the statistically distribution of discharges in historic times remains unknown while the pattern might have been significantly different from recent discharge respectively flood distributions. The significance of the discharge distribution is illustrated by the estimation of the return interval for the catastrophic River Ahr 2021 flood event below.
- Further problems lie in categorising floods according to the damage caused rather than according to hydrological parameters. This can lead to distortions in the assessment of floods, for example if—in a positive case—a flood inundates the fields after they have been harvested, but also if several extreme floods occur in succession without damage having been repaired in the meantime or reconstruction having been possible at all. Note, that no discharge is estimated before a flood frequency analysis is carried out in most cases.
- Furthermore, depending on the data available, floods in long time series have been summarised on a monthly, quarterly, half-yearly or year-round basis, which cannot always be accurately compared with each other due to the time shift. The problem of the identification of the Magdalene Flood in July 1342 in the historic record mentioned above illustrates the challenge.

Despite this abundance of latent and real problems in the evaluation of historical flood series, numerous insights have been gained, which are presented below in a brief review.

Glaser et al. (2010) developed a comparative overview from various existing flood chronologies. The number of floods recorded annually since 1500 was presented as a moving average in order to make it easier to record periods of increased flood frequency and to present them more clearly. Furthermore, the floods were differentiated with regard to their cause in order to be able to show corresponding changes in parallel (Fig. 3.26).

At first glance, the flood frequencies may appear heterogeneous, which is due to the different characteristics and sizes of the catchments of the rivers considered (Glaser et al. 2010). By a closer look, four periods of increased flood frequencies can be identified: 1540–1600, 1640–1700, 1730–1790 and 1790–1840. The first period is particularly pronounced on the Rivers Main and Pegnitz. In the second period, the Danube is added. The third period is less pronounced and can be seen particularly in the Rivers Werra and, to a lesser extent, in the River Main. The fourth phase is characterised by a gradual increase in flood frequencies in all catchments, with all areas subsequently developing differently. This is attributed to the onset of water and river engineering measures, which have had a dominant influence on the floods. Note, that e.g. Blöschl (2022) puts this explanation into perspective for medium size floods only. It is noteworthy that the frequency of floods—i.e. the absolute number of floods in periods of high or low frequency—has not changed significantly. Taking into account flood chronologies of rivers from additional parts of Europe, e.g. Glaser et al. (2010) and Blöschl et al. (2020) confirm the mentioned periods of increased flood frequencies.

As an explanation for the lack of parallelism even in directly neighbouring catchments, the characteristics are assigned to the respective catchments. These include the location and size of lakes (which can buffer flood waves), topographical differences (such as the location of mountain ranges with increased precipitation) and, in particular,

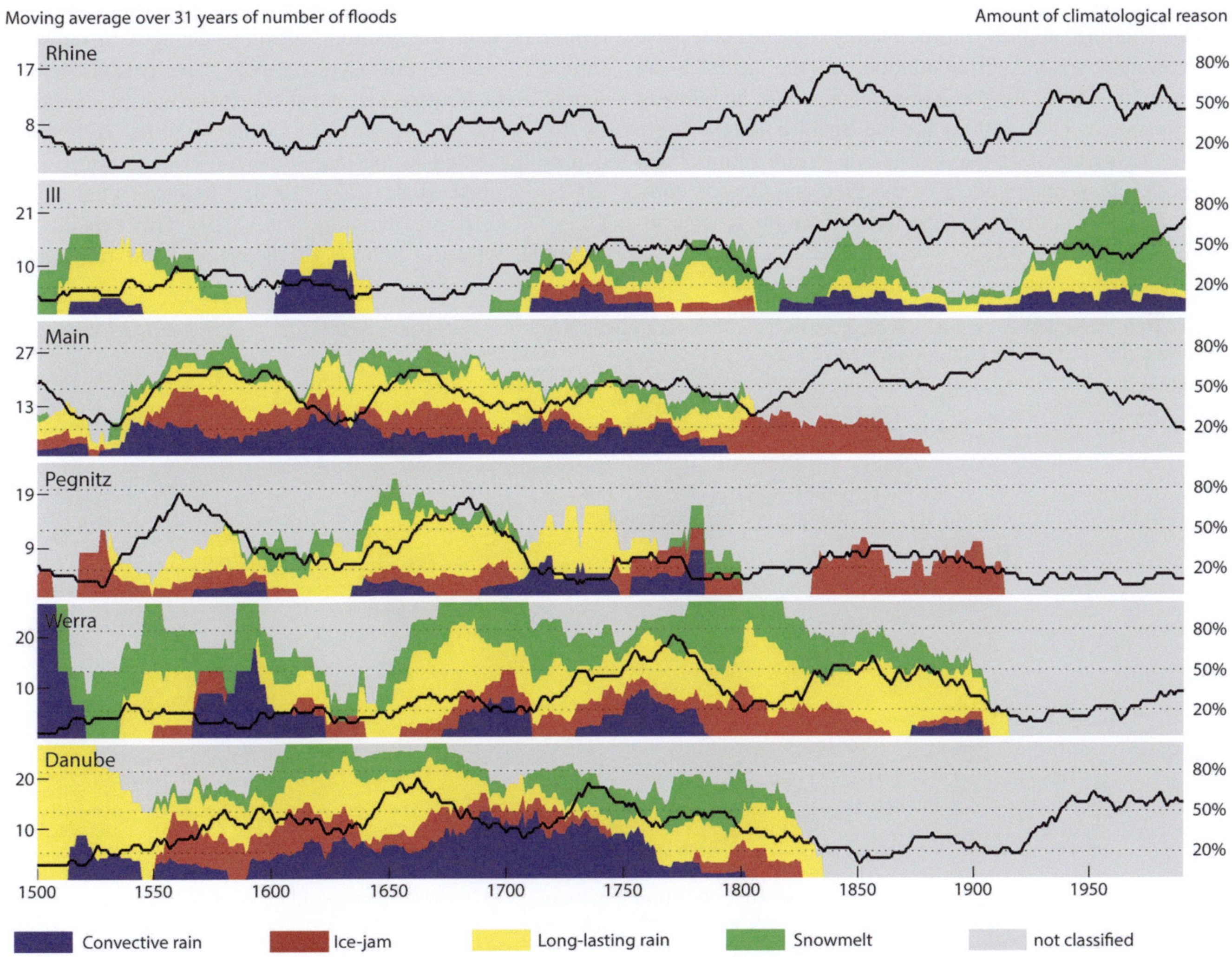

Fig. 3.26 Frequencies and causes of floods since AD 1500 at selected river courses in Central Europe (Glaser et al. 2010, shortened). The black line marks the sum of floods as a moving average over 31 years, while different flood causes are shown stacked in percent by the coloured background

differences in land use affecting runoff (such as the timing and extent of clearing for agricultural use). However, these influencing factors do not act independently of each other, but are coupled with each other. This explains why Mudelsee et al. (2003) could not prove any lasting influence of river straightening, dam construction or deforestation in the flood history of the Rivers Elbe and Oder in the last centuries. However, they can show that for winter floods there is a trend of decreasing flood frequency, which is also attributed to the anthropogenic warming of the rivers and thus the decrease or absence of ice jam floods. In contrast, no change in trend is apparent in the summer floods. Looking back over the past centuries, the River Elbe is in a period of few floods, but the Oder is in a period of numerous floods. However, since the above-mentioned fundamental human influence on the catchment area characteristics affecting runoff should be effective in summer as well as in winter, these factors cannot have been influenced in the long term. Figure 3.27 shows the frequencies of summer and winter floods of the Rivers Elbe and Oder in the last centuries.

This finding applies independently of the otherwise indispensable spatial differentiation, as the analyses by Glaser et al. (2003) for the River Pegnitz near Nuremberg and Barriendos et al. (2003) for the Spanish Rivers Segre and Ter have shown. If the distribution of the summer and winter floods of magnitude 3 of the Elbe and Oder is combined, a similar uniform distribution without obvious periodicities appears (Fig. 3.27).

The picture is somewhat different if one focuses on available gauge data from the last 50 years. In their nationwide analysis, Petrow and Merz (2009) find consistently increasing statistically significant flood trends, whereby the increasing trend is more pronounced for winter floods than for summer floods. On the other hand, Bormann et al. (2011) can show in their evaluation of gauge data that numerous local phenomena can influence the trends of measured values. As usual in statistical analyses, the selection of data as well as the applied methods and assumptions play a central role, as the overview on this by Seidel and Bárdossy (2010) or the controversial discussion on flood trends in Baden-Württemberg between Caspary and Bárdossy (1995) and Vieser (1996) illustrate as example. Nevertheless, Petrow and Merz (2009) confirm with their analysis an increasing risk due to winter floods, while the influence of snowmelt will decrease at least in the middle and lower reaches. Here, at the latest, the importance of changes in atmospheric circulation becomes clear, which is of greater relevance for weather patterns and flood dynamics than the resulting absolute changes in temperature and precipitation (Knox 1993). The periods of different flood frequencies and their trends can be explained by the large-scale circulation, while regional differences can be attributed to the individual tracks (e.g. Sturm et al. 2001; Jacobeit et al. 2003, 2006; Mudelsee et al. 2004; Glaser et al. 2010; Berghuijs et al. 2019). This explains, for example, why the periods of increased flood frequencies

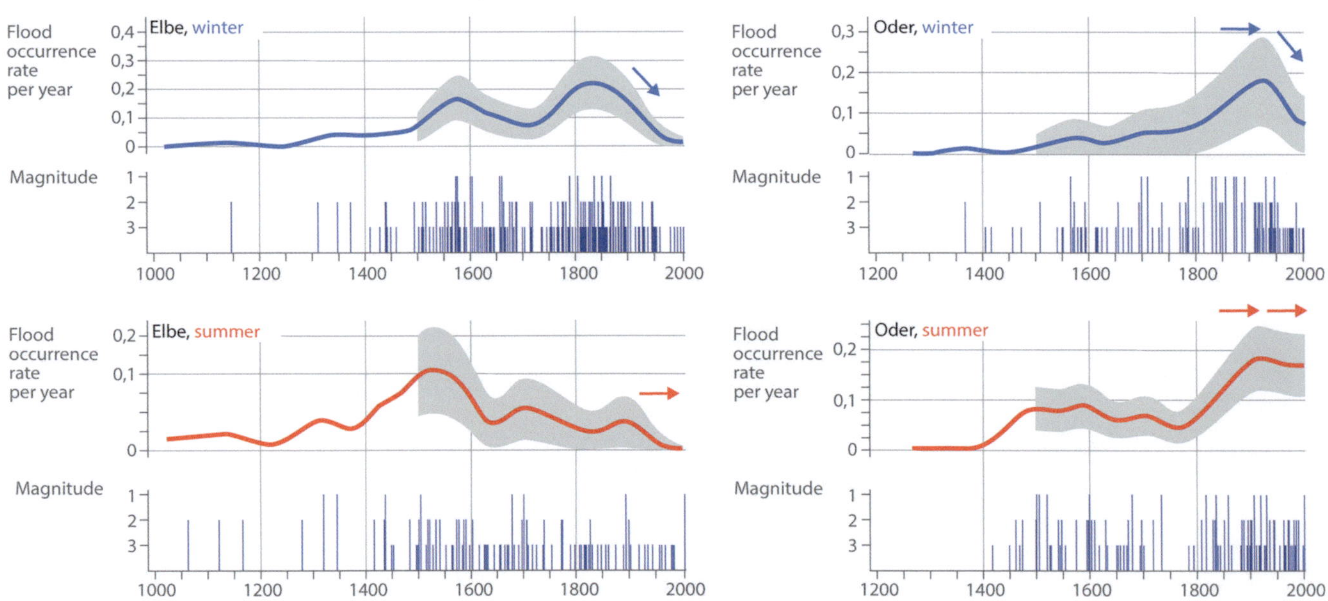

Fig. 3.27 Frequency and magnitude of summer and winter floods of the Rivers Elbe and the Oder (modified after Mudelsee et al. 2003). The number of severe and catastrophic floods (magnitude 2 and 3) has been statistically filtered and converted to annual frequencies. Since it can be assumed that not all floods from the period before 1500 have been recorded, there is no underlaid uncertainty range for this period. The arrows mark the current trends in flood frequencies. It is remarkable that regardless of flood frequencies, which show similarities as well as differences, the temporal distribution of catastrophic extreme events shows no change

specifically in Switzerland correspond very well with the corresponding preceding periods in the rest of Europe, but the correlation of the individual floods between the relatively distant Czech Republic and Switzerland correlates better than with neighbouring Germany: here, regional effects have a stronger impact, while the overarching circulation pattern is identical (Schmocker-Fackel and Naef 2010).

On the other hand—and thereby leaving the focus on Central Europe—contradictory periodicities were frequently found, e.g. in trans-European reviews (e.g. Blöschl et al. 2020) along the Mediterranean Sea (e.g. Camuffo et al. 2002) and China (Fig. 3.16). Obviously, changing the perspective towards a review on a larger spatial scale does not solve the problem of missing plausibility for a specific statistical approach that seems to prove correlations expected as reason to initiate investigations in this context.

This leads to the various attempts to integrate historical flood data into nowadays data sets from gauging stations. The aim is to take additional extreme values into consideration, that are missing in the recorded data set to avoid unexpected large flood magnitudes in the context of risk assessment. Several statistical approaches were developed in this context (e.g. Anderson and Meerschaert 1998; Parkes and Demeritt 2016; Salinas et al. 2016; Lam et al. 2017; St. George and Mudelsee 2019; Fischer and Schumann 2023) frequently considering the fundamental problems of different reasons for flooding and unsteady dynamics due to climate change. Their application to estimate return intervals of extreme flood magnitudes requires various assumptions, e.g. about stochastic data distribution. This is challenging or even doubtful considering climatic change in all temporal perspectives past, recent, and future resulting in constantly changing flood frequencies in both directions simultaneously in neighbouring catchments (e.g. Blöschl et al. 2019). Anyhow, as there is a demand for a number for the return interval, different answers are available as can be illustrated for the catastrophic River Ahr flood in July 2021. Like for high magnitude historic events, the question for the return period of the unexpected flood magnitude of 1000–1200 m^3/s (max. record was 236 m^3/s which is close to HQ_{100}) is estimated in a broad range of values all beyond 1/10,000 years while the last a flood of similar magnitude occurred is just 200 years ago (cf. Roggenkamp and Herget 2024 for details and references).

As already indicated in the introduction of this chapter, there are some obvious problems with flood frequency analysis. A closer look on the concept – beyond mathematics – illustrates this (Baker 1994). Floods events – especially the high magnitude events—are rather individuals and not just members of a homogenous data set and require individual attention.

3.4 Just Before Historic Times—Holocene Flood Events

Since the historical records of flood events only go back a limited time, one has to rely on the analysis of natural flood traces for the preceding periods. These are usually deposits or erosion traces that can be assigned to a flood. Here, a reference to the palaeostage indicators (Chap. 2.2) can be established, although the flood traces do not always have to indicate the exact water level. This is rarely the case for the prehistoric floods considered here, due to the long time since the flood, and if it is, it is more likely to be in orders of magnitude than in centimetre accuracy, as is still frequently known for historic floods. It is important, however, that the traces can be classified in terms of their age, either relatively by their position in relation to other structures or absolutely by dating them with physical methods. A wide range of methods and techniques are available for this purpose, which, however, always place method-specific requirements on the material to be dated (e.g. Reiners et al. 2018). Typically, this is a content of organic material (Fig. 3.28) when using the radiocarbon method or the presence of suitable minerals in the deposited unconsolidated sediments, which also had to be exposed to daylight for a sufficiently long time during transport if a luminescence method was to be used.

In the following, an overview is given to round off the consideration of historical floods, concentrating on the Holocene in Europe, the warm period that has lasted for around 10,000 years since the last glaciation. This time period is of particular interest with regard to the range of fluctuation of the climate parameters, but also the resulting flood

Fig. 3.28 Drilled material from an abandoned meander of the River Lippe valley (Herget 1997). The sedimentary fill of the meander consists of organic sediments that can be dated using the radiocarbon method. Immediately at the bottom are some gravels, followed by brownish sands of the underlying river terrace

parameters, in order to be able to estimate the scope of climate fluctuations within a warm period. References to additional studies are given by e.g. Baker (2008) and Kiss (2019).

Floods that pour into lakes can be clearly identified by changes in the characteristics of the sediments at the bottom of the lake (Wilhelm et al. 2022). Due to the significantly reduced flow velocity when flowing through the lake, the fine-grained suspension load of a river is also deposited during normal discharge. During flood inflow, the higher flow velocity and resulting higher drag force introduces a layer of significantly coarser sediments amidst finer-grained stillwater sediments. Frequent intercalations of such coarse-grained layers can usually be interpreted as indicators of increased flood frequencies. Giguet-Covex et al. (2012), for example, used such an approach in their evidence of increased flood frequencies and magnitudes during the Iron Age and Roman Period in the French Alps. In evaluating lake-bottom landslides attributed to high-energy inflow during floods, Stewart et al. (2010) are able to further show that such relationships do not correlate directly in a linear fashion in terms of frequency. It is also noteworthy that they find fewer flood indicators in Lake Silvaplana in the Upper Engadin for the period 1450 BC to AD 420—i.e. also during the Iron and Roman periods—which is in contrast to the aforementioned study from the French Alps. For this period and the Alps as a whole, a spatial differentiation is obviously necessary. As a third exemplary example for the interpretation of lake sediments with regard to flood dynamics, the investigations of Støren et al. (2010) should be pointed out. For a lake in southern Norway, they were able to reconstruct changes in flood frequencies throughout the entire Holocene, up to historical and recent events. This also includes the legendary "Stor-Ofsen" flood, which is considered the strongest flood in the region in historical times. Its historically recorded intensity provides a benchmark for estimating the magnitude of the prehistoric events since the end of the last glaciation that could be identified so far. This allows not only to delineate periods of different flood frequencies from the lake deposits, but also to determine their magnitude by the grain size and thickness of the individual layers.

The importance of taking prehistoric events into account when estimating flood hazards can be illustrated by Sheffer et al. (2008) in their study of the floods of the Gard River in the Cévennes in southern France. Here, in September 2002, a flood occurred that exceeded all measured and historically recorded events. 23 people died in the floods, 700 houses were damaged and around 1500 houses were submerged by more than 2 m, which illustrates that the building regulations were not coordinated with such flood level. On the basis of datable stillwater deposits, which are located 17–19 m above the mean water level and thus 3 m above that of September 2002, it can be documented that at least five floods have exceeded the extreme event in the last 500 years. Its classification in terms of magnitude is thus clearly put into perspective and the importance of knowledge about prehistoric water levels is evident.

If one goes further back in time beyond the historical periods, it is usually no longer individual flood events that are recorded, but periods of greater flood frequency in general. Due to the imprecision of the dating methods, the temporal resolution is still centuries, which makes the cumulative character obvious. In this context, the floodplain deposits of the river form archives. The flood deposits accumulated there can usually be dated well with the radiocarbon method via their content of organic matter resulting from the erosion of the soils in the catchment area. For example, Hoffmann et al. (2008) were able to prove nine significant periods of increased fluvial activity through statistical analyses of dated floodplain deposits. By comparison with archaeological information on settlement history and especially settlement density, they argue that the peaks of the activity phases around 7050, 6250, 5525, 3690 and 2250 B.C. correlate primarily with fluctuations towards a wetter or cooler climate. Due to increasing settlement density, which is accompanied by agricultural use and resulting soil erosion, the peaks around 1350, 820 and 325 B.C. are no longer dominantly attributable to climatic influence. For the increase in floodplain sediment accumulation since the peak around AD 875, human influence clearly becomes formative. With this different interpretation regarding the cause of the accumulation of floodplain deposits of the same age, it becomes clear that not only an increased number or magnitude of floods comes into question. This indicator can also be influenced in a lasting way in other ways and careful self-critical analyses are indispensable. This is illustrated by the compilation of relevant studies published by Gregory et al. (2006). Here, in addition to other regional studies on hydrological activity phases, important influencing factors (intermediate storage of eroded soil material on slopes; remobilisation from there independent of significant climate changes or flood events) are discussed by examples.

It is remarkable that the evaluation by Hoffmann et al. (2008) presents a supra-regional consistent pattern, because even for the temporally and spatially highly resolved historical period there are parallels in the chronology of flood dynamics. What is striking are the regional but fundamental deviations from the general pattern. How can this be reconciled with a dominant climatic control? Previous approaches to the general chronology of fluvial activity phases have shown more deviations than similarities when examined in detail. Schirmer (1995), for example, in his scheme of three Weichselian and seven Holocene fluvial activity phases admits local and anthropogenic overprinting, just as Hoffmann et al. (2008) do for the more recent phases. If we look at small-scale studies in this field, we find dominant regional-specific factors that hardly show any parallels to an overall scheme. For example, Schielein et al. (2011) illustrate for the valleys of the Rivers Lech and Isar significant influences of the regional runoff regime, Erkens et al. (2009) for the Upper Rhine Graben an overprinting by local tectonic activity, and Herget et al. (2005) a fundamental human impact that began locally much earlier at the turn of the eras due to river engineering measures, which completely suppressed natural dynamics. Consequently, it could be considered whether the catchment areas of the rivers in Central Europe in practice do not rather show individual phases of fluvial activity and stability due to the abundance of natural and anthropogenic factors of influence and whether a supraregional and superordinate pattern only appears subordinate to this. Starkel (2002) also argued in this direction, although he also postulates a general pattern of activity and stability phases.

3.5 Discharge Estimation of Historic Flood Events

As the above explanations have illustrated, information on flood events from historical times has been handed down in varying details. Descriptions of damage are typically the first step in determining the event itself, while information on the duration and the flood level reached form the next higher level. Due to the changes in the channel and the floodplain since historical times as a result of the dredging of rivers to improve navigation and extensive embankment and development of the floodplains, the information cannot be directly transferred to today's conditions. If one wants to derive quantitative statements from historical events beyond the frequency of floods, the discharge must be quantified for these events. For this historical discharge, the resulting water level under today's conditions is then determined via subsequent modelling.

The reconstruction of historical runoff is carried out using the methods presented in Chap. 2, whereby it is necessary to quantify the extent and scope of anthropogenic changes in a spatially and temporally differentiated manner. As this is quite time-consuming, only a few individual studies have been conducted to date. Examples applying different approaches include Wetter et al. (2011) for the River Rhine near Basel, Elleder (2010) and Elleder et al. (2013) for the River Vltava in Prague, Bürger et al. (2006) for the River Neckar in 1824, Calenda et al. (2003) for the 1870 River Tiber flood in Rome, Tetzlaff et al. (2002) for the River Main during the 1342 flood or Schiller (1987), also for the Main (cf. Table 3.1). The sense and necessity of the assumptions and simplifications made in the studies could be debated at length, whereby the focus here is on flood reconstructions within built-up areas. In the following, the reconstruction of the peak discharge of the River Rhine in Cologne during the flood of 1374 illustrates the abundance of parameters that have to be quantified in order to obtain a plausible and comprehensible estimation of the discharge of historical floods.

The reconstruction of the peak discharge of the flood in 1374 is part of a study on the quantification of high magnitude historical floods in Cologne (Herget and Meurs 2010). For this purpose, all floods with a quantifiable water depth of more than 10 m were selected from the available inventories of historical Rhine floods at Cologne. Ice-jam floods were excluded, as due to the backwater effect they are subject to different hydraulic conditions. The available data were compiled on the basis of all source information and references and verified by research in the Cologne city archives, at the Cologne Flood Protection Centre and other regional institutions, thus recording and correcting the surprisingly extensive and diverse traditional errors and inaccuracies (e.g. consideration of the change in the local gauge level zero by − 100 cm in 1979; confusion of events, e.g. with the flood of 1342, numerical errors in data).

The flood of 1374 is particularly noteworthy because, according to contemporary sources, it was possible to cross the city fortification of Cologne by boat, which is a sure indication of a remarkable water level and a traumatic experience for the local population. Since, on the other hand, it is documented that the St. Georgius monastery was not affected by the flood, the maximum water level reached between 12.72 m (cope of the city wall) and 13.52 m (high point at the monastery) in relation to the present-day Cologne gauge level. Comparing and weighting the source texts, as well as taking into account a minimum draught of the boats crossing the city wall, a plausible maximum high-water level of 13.3 m resulted. Unfortunately, the source texts do not allow a clear chronological classification of the flood, so that the precise date of the highest water level is probably 9–11th of February 1374. At Cologne, the flood of winter 1374 lasted about three months with three separate peaks, of which the one in February might be related to the maximum water level (cf. Thelen 1784; Herget and Meurs 2010; Ngo et al. 2023 for details). This was not only the highest flood level without ice-jam, but also the event with the highest volume in the historical flood series of the Rhine in Cologne. Even while the flood occurred in the winter of 1374, the influence of ice floes on the river can be ruled out, because the winter is known to have been extremely warm and had a lot of precipitation (Krahe 1997). The long duration of the flood and the navigation over the city wall during the event also indicate against the influence of ice floes.

In the urban area, a cross-section profile was subsequently determined in which the flow in the event of a flood was not hydraulically disturbed by bridges, weirs, mouths of tributaries, river forks or similar during the period under investigation. Historical descriptions and graphic representations such as engravings and historical maps (e.g. Schwarz 2005) serve as a basis for this. By evaluating all available historical maps and representations, it was also possible to check whether and which maps of more recent times were also suitable for the reconstruction of the topography of 1374. Along the profile line, hydraulically homogeneous units—namely built-up areas, river channel, floodplains and floodplain-channels—were delineated, the flow velocity and discharge components calculated individually, and these subsequently added up.

The peak discharge Q is calculated as a stationary, one-dimensional flow using Manning's equation after insertion into the continuity equation: $Q = A\ R^{2/3}\ S^{1/2}\ n^{-1}$, with A = cross-section area, R = hydraulic radius, S = energy line gradient, n = hydraulic roughness coefficient (cf. Sect. 2.3.1).

The cross-section area A was determined on the basis of the water level (13.3 m + 34.97 m gauge zero = 48.27 m above sea level) using modern topographic data. The changes in relief—generally increases due to sedimentation of alluvial loam and anthropogenic backfilling over the last 500 years—were quantified on the basis of systematic surveys (Heuser and Thielmann 1986) respectively the analysis and interpretation of soil maps. While the amounts in the built-up area vary over small areas and can reach up to several metres, a comparatively low thickness of the alluvial floodplain clay cover of 0.3 m was found for the almost 9 km wide undeveloped floodplain, most of which was only deposited from the eighteenth century onwards. The cross-section profile of the river channel was reconstructed on the basis of a bathymetric survey from the nineteenth century—i.e. before the channel bed was enlarged for the shipping route—and recalculated for the year 1374 using data on incision rates from this period (0.1 m/100 a) (Jasmund 1901). It might be assumed that there were no fundamental changes in the small relief at the channel bottom, which is a negligible inaccuracy against the background of the extraordinarily high water levels of the extreme floods.

The hydraulic radius R as a quotient of the cross-section area A and the wetted perimeter P can be derived from the data of the cross-section area.

In analogy to established simplifications, the energy line gradient S was equated with the water level gradient and it was assumed that since historical times no significant changes in the gradient of the Rhine have occurred during floods within the Cologne urban area. The lowland location of the study area, the absence of barrages or similar backwater structures and the comparatively short river section indicate that the present value of 0.0002 m/m can be transferred to historical times.

The main challenge is the estimation of the hydraulic roughness coefficient n. All factors of influence in general have been taken into account, which can be clearly illustrated in the formula for components (according to Chow 1959):

$$n = (n_1 + n_2 + n_3 + n_4 + n_5 + n_6 + n_7 + n_8 + n_9)\ m,$$

with n_1 = surface roughness, n_2 = vegetation, n_3 = channel bed forms, n_4 = channel bank forms, n_5 = scours and shoals, n_6 = obstacles, n_7 = fluctuating water levels, n_8 = sediment load (density of water), n_9 = seasonal changes and m = correction factor for meander location. For the estimation of the flow velocity during steady flow conditions at peak discharge, not all components are relevant (Herget and Meurs 2010; Herget et al. 2014) and the calculation of the roughness n_p can be simplified to.

$$n_p = (n_1 + n_2 + n_3 + n_4 + n_5 + n_6)\ m$$

The quantification of the components is carried out individually for each element of the cross-section area using the relevant empirical values from the literature (cf. Herget et al. 2014 for details). For the flood in winter 1374, the seasonal reduction of the vegetation density due to missing leafs was taken into account. The ranges of uncertainty respectively missing accuracy of the n-values are captured by considering the minimum n_{min}, maximum n_{max} and plausible mean values n_p and thus ultimately lead to minimum and maximum values of the flow velocity, whereby a mean value can be regarded as a plausible orientation parameter.

Various specific aspects and features can be observed for the different elements of the cross-section area. Observations during the floods of 1993 and 1995 respectively illustrated, that the present-day Cologne urban area with its wide and straight streets compared to historical times can be inundated, but the water in the streets is practically not in motion. This can be explained by the fact that even the streets parallel to the river are not hydraulically continuous, so that they cannot serve as pathway for runoff. An easy indicator are photographs with reflections of buildings on the water surface which are optically sharp only if the water is not moving. The same can be assumed for historical Cologne, which in addition was completely surrounded by the city fortification in 1374, which was also extended towards the river. Therefore, as practically zero flow velocity results in negligible discharge, the built-up areas of Cologne and Deutz can be excluded from the calculation of the peak discharge of historical floods. Against the background of the magnitude of the secondly flood discharge, the volume of the backwatered water in the streets and alleys is undoubtedly negligible.

For the flood in winter 1374, a water level of 48.27 m a.s.l. depending on maximum or minimum roughness coefficients $n_{p\ min}$ and $n_{p\ max}$ results in a peak discharge Q between 18,800 m^3/s and 29,000 m^3/s with a plausible rounded mean discharge of 23,800 m^3/s for mean values of n_p. A differentiation according to the different cross-section elements is shown in Table 3.3.

Table 3.3 Estimated parameters of the 1374 in Cologne (for mean n_p-values of the Manning equation)

	River channel	Floodplain	Floodplain-channels	Entire profile
A—cross-section area (m^2)	4941	7372	3543	15,856
v—mean flow velocity (m/s)	2.73	0.86	1.12	–
Q—peak discharge (m^3/s)	13,489	6340	3968	~23,800

Thus, the peak discharge of this historical flood is, in terms of magnitude, twice as large as the maximum measured flood in Cologne in 1926, which had a peak discharge of 11,100 m^3/s (LUA NRW 2006). The verification of the estimation is carried out by calculating the peak discharge of recent floods using the same approach and comparing it with the available gauge data. For the 1995 flood, taking into account the recent floodplain topography, which is practically limited to the river channel by embankments within the urban area of Cologne, the result was 10,450 m^3/s (calculated) compared to 10,939 m^3/s (measured) (LUA NRW 2006), which corresponds to an underestimation of the peak discharge of almost 5%. Taking into account the considered magnitudes of the flood discharge, the approach can thus be considered as reliable.

Note, that by the application of a 1D-2D coupled hydraulic modelling approach Ngo et al. (2023) estimated a lower peak discharge for the flood event and calculated a return period of about 100,000 years, which fits to the maximum safety standard adopted in the Dutch water policy for some downstream dike sections.

Similar to the procedure in Cologne, estimates of peak discharges of historical floods of the Vltava River were carried out in Prague (Elleder et al. 2013). The investigations showed that the presented approach is not yet universally applicable, because in Prague the flood discharge and the channel bed of the Vltava were already influenced by bridges and weirs in historical times. The reconstruction of the channel bed is difficult because despite the availability of historical bathymetric recordings, no clear tendency of incision or silting up can be determined and extrapolated. Uncertainties were found with regard to the influence of the Charles Bridge on the flow field, as most of the historical flood marks are located directly below the bridge (Fig. 3.12). There, the current is influenced by the bridge piers, which is hardly quantifiable, even though no pressure flow was observed under the bridge during the extreme flood of 2002. The available estimates confirm that not only the water level of the 2002 flood of the Vltava has not been exceeded in historical times (Fig. 3.12), but also the discharge represents an extreme that has never been reached before.

On the other hand, Roggenkamp and Herget (2022) presented first estimates for the peak discharge of the catastrophic flood of River Ahr in western Germany in July 2021 immediately after the event by the application of this approach. The discharge could not be recorded as the gauges along the river were destroyed due to the unexpected large flood magnitude. To avoid miscalculations caused by local backwater effects, flood markers by the uppermost suspension load deposits in sufficient distance from disturbing elements like bridges and buildings were considered. Discharge estimations based on cursory surveys immediately after the event were later confirmed by geodetic surveys at identical locations. A quick estimation was possible as several parameters could be transferred from previous investigations on historic flood events of the river (Roggenkamp and Herget 2014).

As the summary of the approach for estimating the discharge of historical floods shows, the presented method cannot be used for everyday flood estimation. A large number of parameters that differ significantly in small areas, such as floodplain deposit accumulation, channel bed incision or land use, must be quantified with high spatial and temporal resolution, which requires in-depth expertise, practical experiences, and is also quite time-consuming if all parameters have to be determined. A sensitivity analysis based on a sufficient range of data and surveys might help to reduce the time-consuming effort for the data compilation in common practice in the future.

References

Agasse E (2003) Flooding from 17th to 20th centuries in Normandy (western France). In: Thorndycraft VR, Benito G, Barriendos M, Llasat MC (Hrsg.): Palaeofloods, historical data and climatic variability—applications in flood risk assessment. Madrid (Centro de Ciencias Medioambientales), pp 99–107

Aldrete GS (2006) Floods of the Tiber in ancient Rome. Baltimore (John Hopkins University Press)

Alessandroni MG, Remedia G (2002) The most severe flood of the Tiber River in Rome. In: Snorrason A, Finnsdóttir HP, Moss ME (eds) The extremes of the extremes—extraordinary floods. Wallingford (IAHS Press), pp 129–132

Alexandre P (1987) Le climat en Europe au Moyen Age. Paris (École des hautes études en sciences sociales)

Anderson KC, Neff T (2011) The influence of paleofloods on archaeological settlement patterns during AD 1050–1170 along the Colorado River in the Grand Canyon, Arizona, USA. CATENA 85:168–186

Anderson PL, Meerschaert MM (1998) Modeling river flows with heavy tails. Water Resour Res 34:2272–2280

Barrera A, Llasat M-C (2015) Evolving flood patterns in a Mediterranean region (1301–2012) and climatic factors - the case of Catalonia. Hydrol Earth Syst Sci 19:465–483

Barrera A, Llasat M-C, Barriendos M (2006) Estimation of extreme flash flood evolution in Barcelona County from 1351 to 2005. Nat Hazard 6:505–518

Barriendos M, Rodrigo FS (2006) Study of historical flood events of Spanish rivers using documentary data. Hydrolog Sci J 51:765–783
Barriendos M, Martín-Vide J (1998) Secular climate oscillations as indicated by catastrophic floods in the Spanish Mediterranean coastal area (14th–19th centuries). Clim Change 38:473–491
Barriendos M, Coeur D, Lang M, Llasat MC, Naulet R, Lemaitre F, Barrera A (2003) Stationarity analysis of historical flood series in France and Spain (14th–20th centuries). Nat Hazard 2003(3):583–592
Barriendos M, Ruiz-Bellet JL, Tuset J, Mazón J, Balasch JC, Pino D, Ayala JL (2014) The "Prediflood" database of historical floods in Catalonia (NE Iberian Peninsula) AD 1035–2013, and its potential applications in flood analysis. Hydrol Earth Syst Sci 18:4807–4823
Barry JM (1997) Rising tide—the great Mississippi flood of 1927 and how it changed America. New York (Simon and Schuster)
Bartl S, Grünewald U, Schümmberg S (2008) Zur Rekonstruktion langjähriger Wasserstands- und Durchflusszeitreihen am Elbepegel Dresden. Hydrol Wasserbewirtsch 52:162–167
Bauch M (2014) Die Magdalenenflut 1342—ein unterschätztes Jahrtausendereignis? In: Mittelalter-Interdisziplinäre Forschung und Rezeptionsgeschichte. https://mittelalter.hypotheses.org/3016 (27 May 2024)
Bell B (1970) The oldest records of the Nile floods. The Geograph J 136:569–573
Beltaos S (2008) Progress in the study and management of river ice jams. Cold Reg Sci Technol 51:2–19
Belz JU, 14 co-authors (2007) Das Abflussregime des Rheins und seiner Nebenflüsse im 20. Jahrhundert—Analyse, Veränderungen, Trends. Internationale Kommission für die Hydrologie des Rheingebietes KHR, Bericht I-22
Benito G, Díez-Herrero A, Villalta F (2003) Magnitude and frequency of flooding in the Tagus Basin (Central Spain) over the last millennium. Clim Change 58:171–192
Berghuijs WR, Harrigan S, Molnar P, Slater LJ, Kirchner JW (2019) The relative importance of different flood-generating mechanisms across Europe. Water Resour Res 55:4582–4593
BfG = Bundesanstalt für Gewässerkunde (ed) (2002) Das Augusthochwasser im Elbegebiet. Koblenz
Blazer E (1931) Souvenirs montbéliardais, tome 2 - Les inondations à Montbéliard en 1910 et avant. Montbéliard (Ville de Montbéliard).
Blöschl G (2022) Three hypotheses on changing river flood hazards. Hydrol Earth Syst Sci 26:5015–5033
Blöschl G, 41 co-authors (2020) Current flood-rich period is exceptional compared to the past 500 years in Europe. Nature 583:560–566
Blöschl G, 46 co-authors (2019) Changing climate both increases and decreases European river floods. Nature 573:108–111
Böhm O, Wetzel K-F (2006) Flood history of the Danube tributaries Lech and Isar in the Alpine foreland of Germany. Hydrolog Sci J 51:784–798
Böhm O (2011) Hochwassergeschiche des bayerischen Alpenvorlandess. (unpubl. Dissertation, University of Augsburg)
Böhm O, Jacobeit J, Glaser R, Wetzel K-F (2015) Flood sensitivity of the Bavarian Alpine Foreland since the late Middle Ages in the context of internal and external climate forcing factors. Hydrol Earth Syst Sci 19:4721–4734
Bork H-R, Kranz A (2008) Die Jahrtausendflut des Jahres 1342 prägt Deutschland – neue Forschungsergebnisse aus dem Einzugsgebiet des Mains. Jahresberichte der Wetterauischen Gesellschaft Für Die Gesamte Naturkunde 158:119–129
Bork H-R, Russok, C, Dreibrodt S, Dotterweich M, Krabath S, Stephan H-G, Bork H (2006) Spuren des tausendjährigen Niederschlags von 1342. In: Bork H-R (ed) Landschaften der Erde unter dem Einfluss des Menschen. Darmstadt (WBG), pp 115–121
Bork H-R, Bork H, Dalchow C, Faust B, Piorr H-P, Schatz T (1998) Landschaftsentwicklung in Mitteleuropa—Wirkungen des Menschen auf Landschaften. Gotha (Klett-Perthes)
Bormann H, Pinter N, Elfert S (2011) Hydrological signatures of flood trends on German rivers - flood frequencies, flood heights and specific stages. J Hydrol 404:50–66
Börngen M, Tetzlaff G (eds) (2000) Weikinn, C.—Quellentexte zur Witterungsgeschichte Europas von der Zeitenwende bis zum Jahre 1850. Band 5 – 1751 bis 1800. Stuttgart (Borntraeger)
Börngen M, Tetzlaff G (eds) (2002) Weikinn, C.—Quellentexte zur Witterungsgeschichte Europas von der Zeitenwende bis zum Jahre 1850. Band 6 – 1801 bis 1850. Stuttgart (Borntraeger)
Brázdil R (1998) The history of floods in the Rivers Elbe and Vltava in Bohemia. Erfurter Geographische Studien 7:93–108
Brázdil R, Demarée GR, Deutsch M, Garnier E, Kiss A, Luterbacher J, Macdonald N, Rohr C, Dobrovolný P, Kolař P, Chromá K (2010) European floods during the winter 1783/1784 – scenarios of an extreme event during the Little Ice Age. Theoret Appl Climatol 100:163–189
Brázdil R, Dobrovolný P, Elleder L, Kakos V, Kotyza O, Květoň V, Macková J, Müller M, Štekl J, Tolasz, R, Valášek H (2005) Historical and recent floods in the Czech Republic. Brno/Prague (Masaryk University / Czech Hydrometeorological Institut). (in Czech with extended English summary)
Brázdil R, Glaser R, Pfister C, Dobrovolný P, Antoine J-M, Barriendos M, Camuffo D, Deutsch M, Enzi S, Guidoboni E, Kotyza O, Sanchez Rodrigo F (1999) Flood events of selected European rivers in the sixteenth century. Clim Change 43:239–285
Brázdil R, Kundzewicz ZW, Benito G (2006) Historical hydrology for studying flood risk in Europe. Hydrolog Sci J 51:739–764
Brázdil R, Kundzewicz ZW, Benito G, Demarée G, Macdonald N, Roald LA (2012) Historical floods in Europe in the past Millennium. In: Kundzewicz ZW (ed) Changes in Flood Risk in Europe. Wallingford (IAHS), pp 121–166
Brush LM, Wolman MG, Bing-Wei H (eds) (1989) Taming the yellow river—silt and floods. Dordrecht (Kluwer)
Buisman J, van Engelen AFV (1995–2006) Duizend jaar weer wind en water in de Lage Landen. Franeker (Van Wijnen) (5 volumes)
Bürger K, Dostal P, Seidel J, Imbery F, Barriendos M, Mayer H, Glaser R (2006) Hydrometeorological reconstruction of the 1824 flood event in the Neckar River basin (southwest Germany). Hydrol Sci J 51:864–877
Calenda G, Mancini CP, Volpi E (2005) Distribution of the extreme peak floods of the Tiber River from the XV century. Adv Water Resour 28:615–625
Calenda G, Calvani L, Mancini CP (2003) Simulation of the great flood of December 1870 in Rome. Water and Maritime Eng 156:305–312
Camuffo D, Enzi S (1996) The analysis of two bi-millenial series—Tiber and Po River floods. In: Jones PD, RS Bradley, Jouzel J (Hrsg.) Climatic variations and forcing mechanisms of the last 2000 years. Berlin (Springer), pp 433–450
Camuffo D, Sturaro G, Benito G (2002) An opposite flood pattern teleconnection between the Tagus (Iberian Peninsula) and Tiber (Italy) Rivers during the last 1000 years. In: Thorndycraft VR, Benito G (eds) Palaeofloods, historical floods and climatic variability- applications in flood risk assessment. Madrid (CSIC), pp 295–300
Caporali E, Rinaldi M, Casagli N (2005) The Arno River floods. Giornale Di Geologia Applicata 1:177–192
Caspary HJ, Bárdossy A (1995) Markieren die Winterhochwasser 1990 und 1993 das Ende der Stationarität in der Hochwasserhydrologie infolge von Klimaänderungen? Wasser and Boden 47–3:18–24

Champion M (1858–1864) Les inondations en France depuis le VIe siècle jusqu'a nos jours. Paris (Dunod). (6 volumes)

Changming L, Jiyang L (1989) Some problems of flooding and its prevention in the lower Yellow River. In: Brush LM, Wolman MG, Bing-Wei H (eds) Taming the yellow river—silt and floods. Dordrecht (Kluwer), pp 223–241

Cheng Z, Chaogui Z, Chunmei M, Zhibin S, Guangyao Z, Huilin W, Huazhong G, Pengling W, Run H (2005) Identifying paleoflood deposits archived in Zhongba Site, the Three Gorges reservoir region of the Yangtze River, China. Chin Sci Bull 50:2493–2504

Chow VT (1959) Open channel hydraulics. New York (McGraw-Hill)

Clarke AO (1996) Estimating probable maximum floods in the upper Santa Ana Basin, southern California, from stream boulder size. Environ Eng Geosci 2:165–182

Coeur D (2008) La plaine de Grenoble face aux inondations - genèse d'une politique publique du XVIIe au XXe siècle. Versailles (Quae)

Costa JE, Jarrett RD (2008) An evaluation of selected extraordinary floods in the United States reported by the US Geological Survey and implications for future advancement of flood science. US Geological Survey Scientific Investigations Report 2008–5164

Costa JE (1987a) History of paleoflood hydrology in the United States, 1800–1970. In: Landa ER, Ince S (eds) History of hydrology. History of Geophysics 3, Washington (American Geophysical Union), pp 49–53

Costa JE (1987b) Hydraulics and basin morphometry of the largest flash floods in the conterminous United States. J Hydrol 93:313–338

Cyberski J, Grześ M, Gutry-Korycka M, Nachlik E, Kundzewicz ZW (2006) History of floods on the river Vistula. Hydrolog Sci J 51:799–817

De Kraker AMJ (2006) Flood events in the southwestern Netherlands and coastal Belgium, 1400–1953. Hydrolog Sci J 51:913–929

Descombes R (1995) L'eau dans la ville. Des métiers et des hommes. Strasbourg (Hirlé)

Deutsch M, Pörtge K-H (2002) Hochwasserereignisse in Thüringen. Schriftenreihe der Thüringer Landesanstalt für Umwelt und Geologie 63

Deutsch M, Pörtge K-H (2009) Hochwassermarken in Thüringen. Thüringer Ministerium für Landwirtschaft, Forsten, Umwelt und Naturschutz

Deutsch M, Böhner J, Pörtge K-H, Rost KT (2004) Untersuchungen historischer Hochwasserereignisse in Thüringen—dargestellt am Beispiel der Werra. Zeitschrift f. Geomorphologie Suppl 135:11–32

Dion R (1961) Histoire des levees de la Loire. Paris (R. Dion)

DKKV = Deutsches Komitee für Katastrophenvorsorge (ed) (2003) Hochwasservorsorge in Deutschland – Lernen aus der Katastrophe 2002 im Elbegebiet. Bonn.

Dotterweich M, Bork H-R (2007) Jahrtausendflut 1342. Ärchäologie in Deutschland 4(2007):38–40

Dotterweich M, Haberstroh J, Bork H-R (2003) Mittel- und jungholozäne Siedlungsentwicklung, Landnutzung, Bodenbildung und Bodenerosion an einer mittelalterlichen Wüstung bei Friesen, Landkreis Kronach in Oberfranken. Forschungen Zur Deutschen Landeskunde 253:17–56

Duband D (1996) La genèse des crues dans le bassin de la Loire. La Houille Blanche 6–7(1996):54–62

Dutch SI (2009) The largest act of environmental warfare in history. Environ Eng Geosci 15:287–297

Elleder L (2010) Reconstruction of the 1784 flood hydrograph for the Vlatava River in Prague, Czech Republic. Global Planet Change 70:117–124

Elleder L (2015) Historical changes in frequency of extreme floods in Prague. Hydrol Earth Syst Sci 19:4307–4315

Elleder L, Herget J, Roggenkamp T, Nießen A (2013) Historic floods in the city of Prague – reconstruction of peak discharges for 1481–1825 based on documentary sources. Hydrol Res 44:202–214

Ely LL (1997) Response of extreme floods in the southwestern United States to climatic variations in the late Holocene. Geomorphology 19:175–201

Engel H (1999) Eine Hochwasserperiode im Rheingebiet – Extremereignisse zwischen Dez. 1993 und Feb. 1995. Internationale Kommission für die Hydrologie des Rheingebietes KHR, Bericht I-17

Engstrom WN (1996) The California storm of January 1862. Quatern Res 46:141–148

Erkens G, Dambeck R, Volleberg KP, Bouman MTIJ, Bos JAA, Cohen KM, Wallinga J, Hoek WZ (2009) Fluvial terrace formation in the northern Upper Rhine Graben during the last 20 000 years as a result of allogenic controls and autogenic evolution. Geomorphology 103:476–495

Fischer S, Schumann AH (2023) Type-based flood statistics. Cham (Springer)

Francou J, Rodier J-A (1967) Essai de classification des crues maximales observes dans le monde. Cahiers De l'ORSTOM 4:19–46

Fricke W (1988) Der Bericht von E.F. Deurer über das Eishochwasser von 1784—eine Erklärung der Naturkatastrophe. In: Prücker H (ed) Die alte Brücke in Heidelberg. Heidelberg (Braus), pp 41–60

Fu B-J, Wu B-F, Lü Y-H, Xu Z-H, Cao J-H, Niu D, Yang G-S, Zhou Y-M (2010) Three gorges project—efforts and challenges for the environment. Prog Phys Geogr 34:741–754

Fucheng S, Yuanjun Y, Manhua H (1987) Investigation and verification of extraordinarily large floods on the Yellow River. J Hydrol 96:69–78

Fügner D (1995) Hochwasserkatastrophen in Sachsen. Taucha (Tauchaer Verlag)

Fuling X (1989) Evolution of the course of the lower Yellow River in past history. In: Brush LM, Wolman MG, Bing-Wei H (eds) Taming the yellow river—silt and floods. Dordrecht (Kluwer), pp 545–555

Galloway G (2004) USA flood management—Mississippi River—Integrated flood management case study. World Meteorological Organisation

Gees A (1997) Analyse historischer und seltener Hochwasser in der Schweiz. Geographica Bernensia G 53

Georgiadi AG (1993) Historical high water marks as a basis of estimation of spring discharges in Russian Plain rivers. In: Kundzewicz ZW, Rosbjerg D, Simononvic SP, Takeuchi K (Hrsg.): extreme hydrological events—Precipitation, floods and droughts. IAHS Publication vol 213, pp 207–210

Gerhard H (1985) Historische Hochwasser am Oberrhein. Wasserbau-Mitteilungen der TH Darmstadt 24:5–9

Gerlach R (1990) Flußdynamik des Mains unter dem Einfluß des Menschen seit dem Spätmittelalter. Forschungen zur deutschen Landeskunde 234

Ghazi B, Przybylak R, Olinski P, Bogdanska K, Pospieszynska A (2023) The frequency, intensity, and origin of floods in Poland in the 11th–15th centuries based on documentary evidence. J Hydrol 623:129778

Giguet-Covex C, Arnaud F, Enters D, Poulenard J, Millet L, Francus P, David F, Rey P-J, Wilhelm B, Delannoy J-J (2012) Frequency and intensity of high-altitude floods over the last 3.5 ka in northwestern French Alps (Lake Anterne). Quatern Res 77:12–22

Glaser R, Hagedorn H (1990) Die Überschwemmungskatastrophe von 1784 im Maintal – eine Chronologie ihrer witterungsklimatischen Voraussetzungen und Auswirkungen. Erde 121:1–14

Glaser R, Stangl H (2003) Historical floods in the Dutch Rhine Delta. Nat Hazard 3:605–613

Glaser R, Stangl H (2004) Climate and floods in central Europe since AD 1000: data, methods results and consequences. Surveys Geophys 25(5–6):485–510
Glaser R (1991) Klimarekonstruktion für Mainfranken, Bauland und Odenwald anhand direkter und indirekter Witterungsdaten seit 1500. Stuttgart (Fischer)
Glaser R (1998) Historische Hochwässer im Maingebiet – Möglichkeiten und Perspektiven auf der Basis der historischen Klimadatenbank Deutschland (HISKLID). Erfurter Geographische Studien 7:109–128
Glaser R (2008^2) Klimageschichte Mitteleuropas – 1200 Wetter, Klima, Katastrophen. Darmstadt (Wissenschaftliche Buchgesellschaft)
Glaser R, 18 co-authors (2010) The variability of European floods since AD 1500. Climatic Change 101:235–256
Glaser R, Ch. Beck and Stangl H (2003) Zur Temperatur- und Hochwasserentwicklung der letzten 1000 Jahre in Deutschland. In: Deutscher Wetterdienst (ed) Klimastatusbericht 2003, pp 55–67
Görgen K (ed) (2010) Assessment of climate change impacts on discharge in the Rhine River basin—results of the RheinBlick2050 project. Internationale Kommission für die Hydrologie des Rheingebietes KHR, Bericht I-23
Gottschalk MKE (1971–77) Stormvloeden en rivieroverstromingen in Nederland. Assen (van Gorcum) (3 volumes)
Grabs W (ed) (1997) Impact of climate change on hydrological regimes and water resources management in the Rhine basin. Internationale Kommission für die Hydrologie des Rheingebietes KHR, Bericht I-16
Gregory KJ, Macklin MG, Walling DE (eds) (2006) Past hydrological events related to understanding global change. Catena 66:1–187
Grodek T, Benito G, Botero BA, Jacoby Y, Porat N, Haviv I, Cloete G, Enzel Y (2013) The last millennium largest floods in the hyperarid Kuiseb River basin Namib Desert. J Quaternary Sci 28:258–270
Grove J (2004^2) The Little Ice Ages. London (Routledge)
Guidoboni E (1998) Human factors, extreme events and floods in the lower Po plain (northern Italy) in the 16th century. Environ History 4:279–308
Harden T, Macklin MG, Baker VR (2010) Holocene flood histories in south-western USA. Earth Surf Proc Land 35:707–716
Henning R (1904) Katalog bemerkenswerter Witterungsereignisse von den ältesten Zeiten bis zum Jahre 1800. Abhandlungen des Königlich Preussischen Meteorologischen Institutes Band II Nr. 4. Berlin (Asher)
Herget J, Meurs H (2010) Reconstructing peak discharges for historic flood levels in the city of Cologne, Germany. Global Planet Change 70:108–116
Herget J (1997) Die Flußentwicklung des Lippetals. Bochumer Geographische Arbeiten 62
Herget J (2012) Am Anfang war die Sintflut – Hochwasserkatastrophen in der Geschichte. Darmstadt (WBG)
Herget J, Kapala A, Krell M, Rustemeier E, Simmer C, Wyss A (2015) The millennium flood of July 1342 revisted. CATENA 130:82–94
Herget J, Bremer E, Coch T, Dix A, Eggenstein G, Ewald K (2005) Engineering impact on river channels in the River Rhine catchment. Erdkunde 59:Heft 3/4, 294–319
Herschy R (2003) World catalogue of maximum observed floods—IAHS Publication 284. Wallingford (International Association of Hydrological Sciences)
Heuser H, Thielmann G (1986) Ingenieurgeologische Karte 1: 25000 Blatt 5007 Köln. Krefeld (Geologisches Landesamt NRW)
Heuser-Hildebrandt B (2005) Die Hochwasserkatastrophe von 1784 in Mainz – Ursachen, Auswirkungen, Krisenmanagement. Koblenzer Geographisches Kolloquium 27:75–97
Himmelsbach J, Glaser R, Schoenbein J, Riemann D, Martin B (2015) Reconstruction of flood events based on documentary data and transnational flood risk analysis of the Upper Rhine and its French and German tributaries since AD 1480. Hydrol Earth Syst Sci 19:4149–4164
Hjalmarson HW, Phillips JV (1997) Potential effects of translatory waves on estimation of peak flows. J Hydraul Eng 123:571–575
Hjalmarson HW (1984) Flash flood in Tanque Verde Creek, Tucson Arizona. J Hydraulic Eng 110:1841–1852
Hoffmann T, Lang A, Dikau R (2008) Holocene river activity—analysing ^{14}C-dated fluvial and colluvial sediments from Germany. Quatern Sci Rev 27:2031–2040
Horváthová B (2003) Povodeň to nie len vel'ká voda. Bratislava (Veda)
House PK, Pearthree PA (1995) A geomorphologic and hydrologic evaluation of an extraordinary flood discharge estimate—Bronco Creek, Arizona. Water Resour Res 31:3059–3073
House PK, Pearthree PA, Baker VR (1998) Potential effects of translatory waves on estimation of peak flows. J Hydraul Eng 124:1178–1179
Hoyt WG, Langbein WB (1955) Floods. Princeton (Princeton University Press)
IKSE = Internationale Kommission zum Schutz der Elbe (ed) (2004): Dokumentation des Hochwassers vom August 2002 im Einzugsgebiet der Elbe. Magdeburg
Ingram BL, Malamud-Roam F (2012) The West without water—what past floods, droughts, and other climatic clues tell us about tomorrow. Berkeley et al. (University of California Press)
Jacobeit J, Nonnenmacher M, Philipp A (2006) Atmospheric circulation dynamics linked with prominent discharge events in Central Europe. Hydrol Sci J 51:946–965
Jacobeit J, Glaser R, Luterbacher J, Wanner H (2003) Links between flood events in Central Europe since AD 1500 and large-scale atmospheric circulation modes. Geophys Res Lett 30:1172–1175
Jasmund R (1901) Die Arbeiten der Rheinstrom-Bauverwaltung 1851–1900. Halle (Buchdruck des Waisenhauses)
Jiang T, Zhang Q, Blender R, Fraedrich K (2005) Yangtze Delta floods and droughts of the last millennium—abrupt changes and long term memory. Theoret Appl Climatol 82:131–141
Jiaqi C, King L, Tong J, Wollesen D (1998) Klimageschichtliche Forschung in China—Quellenlage und Ergebnisse im Überblick. Erdkunde 52:163–176
Jones RD, Ogilvie AEJ, Wigley TML (1984) Riverflow data for United Kingdom—reconstructed data back to 1844 and historical data back to 1556. Norwich (Climatic Research Unit)
Kale VS (ed) (1998) Flood studies in India. Geological Society of India Memoir 41
Kessler A (1962) Studien zur jüngeren Talgeschichte am Main und an der Mümling und über jüngere Formenentwicklung im hinteren Buntsandstein-Odenwald. Forschungen zur deutschen Landeskunde 133
Kidder TR, Adelsberger KA, Arco LJ, Schilling TM (2008) Basin-scale reconstruction of the geological context of human settlement—an example from the lower Mississippi Valley, USA. Quatern Sci Rev 27:1255–1270
Kington J (1974) The Societas Meteorologica Palatina—an eighteenth century meteorological society. Weather 29:416–426
Kington J (1988) The weather of the 1780s over Europe. Cambridge (Cambridge University Press)
Kiss A (2009) Floods and weather in 1342 and 1343 in the Carpathian Basin. J Environ Geography 2:37–47
Kiss A (2019) Floods and long-term water-level changes in medieval Hungary. Cham (Springer)
Klasz G, Reckendorfer W, Gabriel H, Baumgartner C, Schmalfuss R, Gutknecht D (2014) Natural levee formation along a large and regulated river- the Danube in the National Park Donau-Auen, Austria. Geomorphology 215:20–33
Knox JC (1993) Large increase in flood magnitude in response to modest changes in climate. Nature 361:430–432

Krahe P, Larina M (2010) Hoch- und Niedrigwasser in Köln seit AD 1000. Geographische Rundschau 3 / 2010:34–41

Krahe P (1997) Hochwasser und Klimafluktuationen am Rhein seit dem Mittelalter. In: Immendorf R (ed) Hochwasser – Natur im Überfluß? Heidelberg (Müller), 57–82

Kresser W (1957) Die Hochwässer der Donau. Wien (Springer)

Kuß C (1825) Jahrbuch denkwürdiger Naturereignisse in den Herzogthümern Schleswig und Holstein vom elften bis zum neunzehnten Jahrhundert. Altona (Busch)

Lam D, Thompson C, Croke J, Sharma A, Macklin M (2017) Reducing uncertainty with flood frequency analysis - the contribution of paleoflood and historical flood information. Water Res Res 53. https://doi.org/10.1002/2016WR019959

Lammersen R, Engel H, van de Langemheen W, Buiteveld H (2002) Impact of river training and retention measures on flood peaks along the Rhine. J Hydrol 267:115–124

Lang M, Coeur D (2014) Les inondations remarquables en France—Inventaire 2011 pour la directive Inondation. Versailles (Quae)

Lawler A (2010) Uncovering a rural Chinese Pompeii. Science 328:566–567

LfU = Bayerisches Landesamt für Umwelt (ed) (2008) Leben mit dem Fluss - Hochwasser im Spiegel der Zeit. Augsburg (LfU).

Llasat MC, Barriendos M, Barrera A, Rigo T (2005) Floods in Catalonia (NE Spain) since the 14th century. J Hydrology 313:32–47

LUA NRW = Landesumweltamt Nordrhein-Westfalen (ed) (2006) Deutsches Gewässerkundliches Jahrbuch Rheingebiet III 2003. Essen (Landesumweltamt NRW)

LUMN = Landesamt für Umwelt, Messungen und Naturschutz Baden-Württemberg (2006) Historische Hochwassermarken in Baden-Württemberg (CD-ROM). Karlsruhe

Luo C-Z (1987) Investigation and regionalization of historical floods in China. J Hydrology 96:41–51

Macdonald N, Black AR (2010) Reassessment of flood frequency using historical information for the River Ouse at York, UK (1200–2000). Hydrol Sci J 55:1152–1162

Macdonald N, Sangster H (2017) High-magnitude flooding across Britain since AD 1750. Hydrol Earth Syst Sci 21:1631–1650

MacDonald N (2013) Reassessing flood frequency for the River Trent through the inclusion of historical flood information since AD 1320. Hydrol Res 44:215–233

MacDonald N et al. (2006) Historical and pooled flood frequency analysis for the River Tay at Perth, Scotland. Area 38:34–46

Machado MJ, Benito G, Barriendos M, Rodrigo FS (2011) 500 years of rainfall variability and extreme hydrological events in southeastern Spain drylands. J Arid Environ 75:1244–1253

McEwen L (1987) Sources for establishing a historic flood chronology (pre-1940) within Scottish River Catchments. Scott Geogr Mag 103:132–140

Milliman JD, Farnsworth KL (2011) River discharge to the coastal ocean—a global synthesis. Cambridge University Press, Cambridge

Morrill CJT, Overpeck & J.E. Cole, (2003) A synthesis of abrupt changes in the Asian summer monsoon since the last deglaciation. Holocene 13:465–476

Mudelsee M, Börngen M, Tetzlaff G (2006) Trends in flood risk of the River Werra (Germany) over the past 500 years. Hydrological Sci J 51:818–833

Mudelsee M, Börngen M, Tetzlaff G, Grünewald U (2003) No upward trends in the occurence of extreme floods in Central Europe. Nature 425:166–169

Mudelsee M, Börngen M, Tetzlaff G, Grünewald U (2004) Extreme floods in Central Europe over the past 500 years - role of cyclone pathway "Zugstrasse Vb." J Geophys Res 109:D23101

Munoz SE, Giosan L, Therrell MD, Remo JWF, Shen Z, Sullivan RM, Wiman C, O'Donnell M, Donnelly JP (2018) Climatic control of Mississippi River flood hazard amplified by river engineering. Nature 556:95–98

Munzar J, Elleder L, Deutsch M (2005) The catastrophic flood in February/March 1784 – a natural disaster of European scope. Moravian Geograph Rep 13:8–24

Naulet R, Lang M, Ouarda TBMJ, Coeur D, Bobée B, Recking A, Moussay D (2005) Flood frequency analysis on the Ardèche river using French documentary sources from the last two centuries. J. of Hydrology 313:58–78

Negm AM (ed) (2020) Flash floods in Egypt. Cham (Springer)

Ngo H, Bomers A, Augustijn DCM, Ranasinghe R, Filatova T, van der Meulen B, Herget J, Hulscher SJMH (2023) Reconstruction of the 1374 Rhine river flood event around Cologne region using 1D-2D coupled hydraulic modelling approach. J Hydrol 617. https://doi.org/10.1016/j.jhydrol.2022.129039

O'Connor JE, Costa JE (2003) Large floods in the United States—where they happen and why. US Geological Survey Circular 1245

O'Connor JE, Costa JE (2004) The world's largest floods, past and present—their causes and magnitudes. US Geological Survey Circular 1254

O'Connor JE, Grant GE, Costa JE (2002) The geology and geography of floods. In: House PK, Webb RH, Baker VR, Levish DR (Hrsg.) Ancient floods, modern hazards—principles and applications of palaeoflood hydrology. Washington (American Geophysical Union), pp 359–385

Pang KD (1987) Extraordinary floods in early Chinese history and their absolute dates. J Hydrol 96:139–155

Paprotny D, Sebastian A, Morales-Nápoles O, Jonkman SN (2018) Trends in flood losses in Europe over the past 150 years. Nat Commun 9:1985

Parde M (1961) Sur la puissance des crues en diverses parties du monde. Zaragoza (Departamento de geografía aplicada del Instituto Elcano)

Parkes BL, Demeritt D (2016) Defining the hundred-year flood—a Bayesion approach for using historic data to reduce uncertainty in flood frequency estimates. J Hydrol 540:1189–1208

Pavese MP, Banzon V, Colacino M, Gregori GP, Pasqua M (1992) Three historical data series on floods and anomalous climatic events in Italy. In Bradley RS, Jones PD (eds) Climate since AD 1500, London (Routledge), pp 155–170

Pekárová P, Halmová D, Bačová Mitková V, Miklánek P, Pekár J, Škoda P (2013) Historic flood marks and flood frequency analysis of the Danube River at Bratislava, Slovakia. J Hydrol Hydromech 61:326–333

Petrow T, Merz B (2009) Trends in flood magnitude, frequency and seasonality in Germany in the period 1951–2002. J Hydrol 371:129–141

Pfister C, Wanner H (2021) Klima und Gesellschaft in Europa. Bern (Haupt)

Pfister C, Hächler S (1991) Überschwemmungskatastrophen im Schweizer Alpenraum seit dem Spätmittelalter. Würzburger Geograph Arbeiten 80:127–148

Pfister C (1999) Wetternachhersage—500 Jahre Klimavariationen und Naturkatastrophen. Bern (Haupt)

Pfister C (2006) Überschwemmungen und Niedrigwasser im Einzugsgebiet des Rheins 1500–2005. In: Bürgin T, Capaul U, Hilfiker H, Klötzli F, Müller JP, Schläfli A (Hrsg.) Der Rhein—Lebensader einer Region. Neujahrsblatt der Naturforschenden Gesellschaft Zürich auf das Jahr 2006 (208):265–273

Pichard G, Roucaute E (2014) Sept siècles d'histoire hydroclimatique du Rhône d'Orange à la mer (1300–2000) climat, crues, inondations. Méditerranée Horssérie 192

Pichard G, Arnaud-Fassetta G, Moron V, Roucaute E (2017) Hydroclimatology of the Lower Rhône Valley: historical flood reconstruction (AD 1300–2000) based on documentary and instrumental sources. Hydrol Sci J 62:1772–1795

Pierik HJ, Stouthamer E, Schuring T, Cohen KM (2018) Human-caused avulsion in the Rhine-Meuse delta before historic embankment, The Netherlands. Geology 46:935–938

Pohl R (2004a) Das Sommerhochwasser 2002 in Dresden und Umgebung. Erfurter Geographische Studien 11:121–128

Pohl R (2004b) Historische Hochwasser aus dem Erzgebirge. Dresdner Wasserbauliche Mitteilungen 28

Popper W (1951) The cairo nilometer. University of California Press, Berkeley

Porter K, 39 co-authors (2011) Overview of the ARkStorm scenario. US Geological Survey Open-File Report 2010–1312

Pötzsch CG (1784) Chronologische Geschichte der großen Wasserfluthen des Elbstroms seit tausend und mehr Jahren. Dresden (Hofbuchhandlung Walther)

Pötzsch CG (1786) Chronologische Geschichte der großen Wasserfluthen des Elbstroms seit tausend und mehr Jahren—Nachtrag und Fortsetzung. Dresden (Hofbuchhandlung Walther)

Prieto MR (2007) ENSO signals in South America – rains and floods in the Paraná River region during colonial times. Clim Change 83:39–54

Qingchao Y (1989) Fluvial processes of the lower Yellow River and estimation of flood conditions. In: Brush LM, Wolman MG, Bing-Wei H (eds) Taming the Yellow River – silt and floods. Dordrecht (Kluwer), pp 261–274

Reiners PW, Carlson RW, Renne PR, Cooper KM, Granger DE, McLean NM, Schoen B (2018) Geochronology and thermochronology. Chichester (Wiley)

Retsö D (2015) Documentary evidence of historical floods and extreme rainfall events in Sweden 1400–1800. Hydrol Earth Syst Sci 19:2561–2576

Roald LA (2021) Floods in Norway—NVE Rapport 1/2021. Oslo (Norges vassdrags- og energidirektorat)

Röckel D (1994) Der Neckar und seine Hochwasser am Beispiel von Eberbach. Eberbach (Wilhelm Krauth)

Rodier JA, Roche M (1984) World catalogue of maximum observed floods—IAHS Publication 143. Wallingford (International Association of Hydrological Sciences)

Roggenkamp T, Herget J (2014) Reconstructing peak discharges of historic floods of the River Ahr, Germany. Erdkunde 68:49–59

Roggenkamp T, Herget J (2022) Hochwasser der Ahr im Juli 2021—Abflusseinschätzung und Einordnung. HyWa – Hydrologie und Wasserbewirtschaftung 66:40–49

Roggenkamp T, Herget J (2024) Flood reconstruction—the unexpected rather frequent event at River Ahr in July 2021. Global and Planetary Change. https://doi.org/10.1016/j.gloplacha.2024.104541

Rohr C (2003) Der Fluss als Ernährer und Zerstörer. Zur Wahrnehmung, Deutung und Bewältigung von Überschwemmungen an den Flüssen Salzach und Inn (13.-16. Jahrhundert). Traverse 10/3, 37–49

Rohr C (2005) The danube floods and their human response and perception (14th to 17th C). History Meteorol 2:71–86

Rohr C (2007) Extreme Naturereignisse im Ostalpenraum—Naturerfahrung im Spätmittelalter und am Beginn der Neuzeit. Wien (Böhler)

Rohr C (2010) Measuring the frequency and intensity of floods of the River Traun (Upper Austria) 1441–1574. Hydrol Sci J 51:834–847

Roth R (1996) Einige Bemerkungen zur Entstehung von Sommerhochwasser aus meteorologischer Sicht. Zeitschrift Für Kulturtechnik u. Landentwicklung 37:241–245

Röthlisberger G (1991) Chronik der Unwetterschäden in der Schweiz. Berichte der Eidg. Forschungsanstalt für Wald, Schnee u. Landschaft 330

Rutte E, Wilczewski N (1983^3) Mainfranken und Rhön. Sammlung Geologischer Führer 74. Stuttgart (Borntraeger)

Said R (1993) The River Nile—geology, hydrology and utilization. Oxford (Pergamon)

Saito Y, Yang Z, Hori K (2001) The Huanghe and Changjiang deltas—a review on their characteristics, evolution and sediment discharge during the Holocene. Geomorphology 41:219–231

Salinas JL, Kiss A, Viglione A, Viertl R, Blöschl G (2016) A fuzzy Bayesian approach to flood frequency estimation with imprecise historical information. Water Resour Res 52:6730–6750

Sartor J (2012) Historische Hochwasser der Mosel. Jahrbuch Bernkastel-Wittlich 2012:244–249

Schielein R, Schellmann G, Lomax J (2011) Stratigraphy of late quaternary fluvial terraces at the confluence of the lech and danube valleys. E&G Quaternary Science Journal 60:414–424

Schiller H (1987) Ermittlung von Hochwasserwahrscheinlichkeiten am schiffbaren Main und überregionaler Vergleich der Ergebnisse. Beiträge Zur Hydrologie S6:79–101

Schirmer W (1995) Valley bottoms in the late Quaternary. In: Hagedorn J (ed) Late Quaternary and present-day fluvial processes in Central Europe. Zeitschrift für Geomorphologie N.F. Suppl.-Bd. 100:27–51

Schmidt M (2000) Hochwasser und Hochwasserschutz in Deutschland vor 1850. München (Oldenbourg)

Schmidt M (2002) Historische Hochwasser Im Deutschen Rheingebiet. Wasserwirtschaft 4–5(2002):48–52

Schmocker-Fackel P, Naef F (2010) Changes in flood frequencies in Switzerland since 1500. Hydrol Earth Syst Sci 14:1581–1594

Schneider H (2001) Überschwemmungen und Hochwasserschutz im antiken Rom. Kasseler Wasserbau Mitteilungen 12:203–207

Schulte L, Peña JC, Carvalho F, Schmidt TL, Ramon JB, Llorca J, Veit H (2015) A 2600-year history of floods in the Bernese Alps, Switzerland: frequencies, mechanism and climate forcing. Hydrol Earth Syst Sci 19:3047–3072

Schwarz U (ed) (2005) Köln und sein Umland in alten Karten. Köln (Emons)

Seberich F (1958) Die alte Mainbrücke zu Würzburg. Mainfränkische Hefte 31

Seidel J, Bárdossy A (2010) Berücksichtigung von historischen Hochwasserereignissen in der Extremwertstatistik. Geographische Rundschau 3:42–47

Sheffer NA, Rico M, Enzel Y, Bernito G, Grodek T (2008) The palaeoflood record of the Gardon River, France - a comparison with the extreme 2002 flood event. Geomorphology 98:71–83

Sidler WA (1968) Agua Mansa and the flood of January 22, 1862, Santa Ana River. San Bernadino County Flood Control District

Siegenthaler C, Sturm M (1991) Die Häufigkeit von Ablagerungen extremer Reuss-Hochwasser - die Sedimentationsgeschichte im Urnersee seit dem Mittelalter. Mitteilungen Des Bundesamtes Für Wasserwirtschaft 4:127–139

Smith DT, Reed DB (1990) A centennial survey of American floods—fifteen significant events in the United States 1890–1990. NOAA Technical Memorandum NWS SR-133

Snorasson A, Finnsdóttir HP, M Moss (Hrsg.) (2002) The extremes of the extremes—extraordinary floods. IAHS Publication 271

St. George S, Mudelsee M (2019) The weight of the flood-of-record in flood frequency analysis. J Flood Risk Manage 12 e12512

Starkel L (2002) Change in the frequency of extreme events as the indicator of climatic change in the Holocene (in fluvial systems). Quatern Int 91:25–32

Stewart MM, Grosjean M, Kuglitsch FG, Nussbaumer SU, von Gunten L (2010) Reconstructions of late Holocene paleofloods and glacier length changes in the Upper Engadine, Switzerland (ca. 1450 BC -AD 420). Palaeogeogr Palaeoclimatol Palaeoecol 311:215–223

Støren EN, Dahl SO, Nesje A, Paasche Ø (2010) Identifying the sedimentary imprint of high-frequency Holocene river floods in lake sediments - development and application of a new method. Quatern Sci Rev 29:3021–3033

Sturm K, Glaser R, Jacobeit J, Deutsch M, Brázdil R, Pfister C (2001) Hochwasser in Mitteleuropa seit 1500 und ihre Beziehung zur atmosphärischen Zirkulation. Petermanns Geogr Mitt 145:18–27

Sutcliff JV, Parks YP (1999) The hydrology of the Nile. IAHS Special Publication 5

Svoboda A, Pekárová P, Miklánek P (2000) Flood hydrology of the Danube between Devín and Nagymaros. Bratislava (Institute of Hydrology and Slovak National Committee on Hydrology)

Tetzlaff G, Börngen M, Mudelsee M, Raabe A (2002) Das Jahrtausendhochwasser von 1342 am Main aus meteorologisch-hydrologischer Sicht. Wasser and Boden 54(10):41–49

Thelen JL (1784) Ausführliche Nachricht von dem erschrecklichen Eisgange, und den Überschwemmungen des Rheines, welche im Jahre 1784 die Stadt Köln, und die umliegenden Gegenden getroffen. Köln

Thordarson T, Self S (2003) Atmospheric and environmental effects of the 1783–1784 Laki eruption—a review and reassessment. J Geophys Res D1-108. https://doi.org/10.1029/2001JD002042

Thorndycraft VR, Benito G, Barriendos M, Llasat (Hrsg.) MC (2003) Palaeofloods, historical data & climatic variability – applications in flood risk assessment. Madrid (Centro de Ciencias Medioambientales)

Tol RSJ, Langen A (2000) A concise history of Dutch river floods. Clim Change 46:357–369

Toonen WHJ (2015) Flood frequency analysis and discussion of non-stationarity of the Lower Rhine flooding regime (AD 1350–2011): Using discharge data, water level measurements, and historical records. J Hydrol 528:490–502

USGS = US Geological Survey (ed) (1995) Floods in the upper Mississippi River basin, 1993. US Geological Survey Circular 1120

USGS = US Geological Survey (ed) (2013) 2011 Floods of the central United States. US Geological Survey Professional Paper 1798

Valdés-Manzanilla A (2016) Historical floods in Tabasco and Chiapas during sixteenth-twentieth centuries. Nat Hazards 80:1563–1577

van der Meulen B (2021) Flood marks on buildings. Utrecht Stud Earth Sci 236:45–71

Vieser H (1996) Stationarität historischer Reihen extremer Hochwasser in Baden-Württemberg. Wasser & Boden 48(8):50–54

Vischer DL (2003) Die Geschichte des Hochwasserschutzes in der Schweiz—von den Anfängen bis ins 19. Jahrhundert. Berichte des BWG (Bern) Serie Wasser 5

Wallner J (1957) Hochwasserabfluß und 100 Jahre wasserbauliche Maßnahmen—eine Untersuchung über den 400 km langen schiffbaren Main von Bamberg bis zur Mündung. Besondere Mitteilungen zum Deutschen Gewässerkundlichen Jahrbuch 20

Weber H (1977) Hochwasser Im Rheingebiet. Beiträge Zur Rheinkunde 29:50–62

Wehnert H, Paczkowski J (1985) Hochwasser in Wertheim. Wertheim (Wehnert)

Weikinn C (1958) Quellentexte zur Witterungsgeschichte Europas von der Zeitenwende bis zum Jahre 1850. Band 1 – Zeitenwende bis 1500. Berlin (Akademie)

Weikinn C (1960) Quellentexte zur Witterungsgeschichte Europas von der Zeitenwende bis zum Jahre 1850. Band 2 – 1501 bis 1600. Berlin (Akademie)

Weikinn C (1961) Quellentexte zur Witterungsgeschichte Europas von der Zeitenwende bis zum Jahre 1850. Band 3 – 1601 bis 1700. Berlin (Akademie)

Weikinn C (1963) Quellentexte zur Witterungsgeschichte Europas von der Zeitenwende bis zum Jahre 1850. Band 4 – 1701 bis 1750. Berlin (Akademie)

Wetter O, Pfister C, Weingartner R, Luterbacher J, Reist T, Trösch J (2011) The largest floods in the High Rhine basin since 1268 assessed from documentary and instrumental evidence. Hydrolog Sci J 56:733–758

Wilhelm B, Amann B, Corella JP, Rapuc W, Giguet-Covex C, Merz B, Storen E (2022) Reconstructing paleoflood occurrence and magnitude from lake sediments. Quaternary 5. https://doi.org/10.3390/quat5010009

Williams A, Archer D (2002) The use of historical flood information in the English Midlands to improve risk assessment. Hydrolog Sci J 47:67–76

Witte W, Krahe P, Liebscher H-J (1995) Rekonstruktion der Witterungsverhältnisse im Mittelrheingebiet von 1000 n.Chr. bis heute anhand historischer hydrologischer Ereignisse. Internationale Kommission für die Hydrologie des Rheingebietes KHR, Bericht II-9

Woodward JC, Macklin MG, Krom MD, Williams MAJ (2007) The Nile—evolution, quaternary river environments and material fluxes. In: Gupta A (ed) Large rivers—geomorphology and management. Chichester (Wiley), pp 261–292

Xu J (2003) Sedimentation rates in the lower Yellow River over the past 2300 years as influenced by human activities and climate change. Hydrol Process 17:3359–3371

Xue C (1993) Historical changes in the Yellow River delta, China. Mar Geol 113:321–329

Yang D, Yu G, Xie Y, Zhan D, Li Z (2000) Sedimentary records of large Holocene floods from the middle reaches of the Yellow River, China. Geomorphology 33:73–88

Yeo SW, Blong RJ, McAneny KJ (2007) Flooding in Fiji – findings from a 100-year historical series. Hydrolog Sci J 52:1004–1015

Yin H, Li C (2001) Human impact on floods and flood disasters on the Yangtze River. Geomorphology 41:105–109

Zbinden E (2011) Das Magdelenen-Hochwasser von 1342 - der "hydrologische Gau" in Mitteleuropa. Wasser Energie Luft 103:193–203

Zhang L (2009) Changing with the Yellow River - an environmental history of Hebei, 1048–1128. Havard J Asiatic Stud 69:1–36

Zhang Q, Gemmer M, Chen J (2008) Climate changes and flood/drought risk in the Yangtze Delta, China, during the past millennium. Quatern Int 176(177):62–69

Further Reading

Reader without experiences in statistics might benefit from the elementary introduction prepared by the US Geological Survey:

England JF, Cohn TA, Faber BA, Stedinger JR, Thomas WO, Veilleux AG, Kiang JE, Mason RR (2018) Guidelines for determining flood flow frequency - Bulletin 17C. US Geological Survey Techniques and Methods, book 4, chap. B5, https://doi.org/10.3133/tm4B5

A critical perspective towards flood frequency analysis is presented by

Baker VR (1994) Geomorphological understanding of floods. Geomorphology 10:139–156

References on additional studies of flood reconstructions of Holocene times are provided e.g. by:

Kiss A (2019) Floods and long-term water-level changes in medieval Hungary. Cham (Springer)

Baker VR (2008) Paleoflood hydrology – origin, progress, prospects. Geomorphology 101:1–13

A review on the method for estimation of the discharge of historic flood levels presented including a discussion of several considerations and references is given in

Herget J, Roggenkamp T, Krell M (2014) Estimation of peak discharges of historical floods. Hydrol Earth Syst Sci 18:4029–4037

4 Another Process Pattern: Outburst Floods

In contrast to floods, which are controlled by meteorological causes such as precipitation or snowmelt, lakes can empty suddenly and unexpectedly without warning or external signs. Depending on the volume of water released, extreme peak discharges can occur that exceed floods caused by precipitation by several orders of magnitude. Although reservoirs may be associated primarily with man-made dams, there are many forms of natural reservoirs whose abrupt emptying due to dam failure is referred to as a lake outburst.

Since the different types of reservoirs have fundamentally different dynamics, an overview of the different types of reservoirs and the spectrum of relevant outburst mechanisms is first given. In order to be able to better imagine and understand the processes in the past, recent reservoirs and outburst events are also included in this consideration. Subsequently, the spectrum of outburst floods and their consequences is presented by selected examples in further detail.

4.1 Naturally Dammed Lakes

There is a wealth of reviews on reservoirs and the outburst floods they generate (including Costa 1988; Costa and Schuster 1988, Cenderelli 2000; O'Connor et al. 2002, Korup and Tweed 2007; O'Connor and Beebee 2009; Liu et al. 2019), which also consider those of natural origin in particular and can be differentiated as follows in a first overview.

Water collects in **basins and depressions** following gravity and can evaporate from there or flow out through spillways and outlets. These basins can be formed by tectonic movements of the earth's crust or erosion processes on the earth's surface and are filled up over time by sediments. This silting up is the common pattern for a lake, so that one would hardly think of seeing damming effects and outburst potentials here. However, if the water balance in the catchment area changes, for example due to climate change, the water level can rise significantly higher than usual. Local watersheds can be overcome and water can flow into previously dry valleys. As a result of retrograde erosion at this overflow, the spillway can incise considerably and large quantities of water originally draining or evaporating elsewhere can be released. The dimension of the remaining gap hinders the refill of the base, so the outburst flood is a one-time only event. A prominent example of such a lake overflow is the emptying of Lake Bonneville in the area of today's Great Salt Lake in the US state of Utah, which is described in more detail in the context of Pleistocene megafloods (cf. Sect. 5.3).

Gravitational mass movements such as rockfalls and landslides are typically triggered by heavy rainfall or earthquakes. The rock masses displaced by gravity can fill valley floors to varying degrees (which is the starting point for a more extensive classification by Costa and Schuster 1988) and dam up the river courses there to form lakes. Obviously, the stability and durability of such a dam depends on the ratio of the amount of displaced rock to the inflow into the lake. Since there is no consolidation and stabilisation of the dam material, in practice such dams are either destroyed comparatively quickly, usually by overflow, or last for a very long time, as in the example from the Tien-Shan Mountains shown in Fig. 4.1.

Ice dams caused by valley glaciers or ice sheets can block the course of a river, similar to the deposits of gravitational mass movements, and thus lead to the damming of a lake. This results in a multifaceted interplay, because liquid water can, on the one hand, melt ice and even melt it completely, it allows ice to float, as drifting ice floes and icebergs show, but it is also dammed by it, as the various ice-dammed lakes illustrate. The gradual and only temporary backwatering and damming of river water during ice-jam floods by rising ice floes was illustrated above using the example of the historical flood of 1784. Liquid water can be impounded by ice on the surface in different spatial

J. Herget, *Flood Reconstruction*, Springer Natural Hazards, https://doi.org/10.1007/978-3-031-66815-9_4

Fig. 4.1 Upper and Lower Ikhnach Lake in the Tien-Shan Mountains, Uzbekistan. On the opposite valley slope, the landslide which dam the upper lake is clearly visible

configurations, but in addition it can also form lakes on, in and under the ice, which are referred to as supra-, en- and subglacial reservoirs. Typically, the outflow of streams in tributary valleys that are dammed by a large valley glacier in the main valley is blocked. Figure 4.2 illustrates this with an example from Iceland.

Due to the numerous ice dammed lake outburst floods in Iceland (Carrivick and Tweed 2019), where many volcanoes carry glaciers or are even covered by them and therefore produce large amounts of subglacial meltwater during eruptions, the Icelandic term "jökulhlaup" (glacial flood) is used in the international literature for these sudden and intense outpourings of water from the ice. The tendency is for the number of jökulhlaups to decrease as the glaciers melt. However, it must be taken into account that it is not the large ice masses that are responsible for the formation of dams, but rather the spatial configuration of the glaciers into river courses and their internal structure, which can also back up and abruptly release glacial meltwater.

After thawing and retreat of the active ice margin, **terminal and lateral moraines** remain at the front and sides of former glacier margins. In combination with the basin

Fig. 4.2 Lake Graenalon in Iceland is dammed by the extensive ice sheet of Vatnajökull, which reaches to the horizon

Fig. 4.3 Lake Steinsee at Susten Pass, Switzerland. The residual lake is bounded by the terminal moraine of 1920, after a previously higher lake had drained. Back-cutting along the spillway caused the lake to drain in July 1956, which also created the prominent gully below the lake (King 1987)

deepened by exaration of the glacier, they literally form the framework for corresponding reservoirs fed by the melt water of the glacier as well as precipitation (Fig. 4.3).

With increasing age, moraine dams tend to become unstable. This is primarily due to thawing ice remnants that remained covered by deposits as so-called dead ice when the glacier thawed. In addition, thawing permafrost occurs when the moraine deposits were stabilised by freezing and lose their cohesion through warming of this binding agent. The sediments, which may originally have been stored in an over-steep manner, follow gravity, collapse or flatten due to slides, causing the moraine wall to become irregularly lowered. **Thermokarst lakes**, which form in depressions in the permafrost, can also be named in this context. The water melts the frozen subsoil and the shore areas, so that the lake grows and several lake basins might join together.

As a result of **volcanic eruptions,** natural reservoirs of different types can develop. The outflowed lava and the ejected pyroclastics can divert river courses and dam them up to form lakes. Loosely deposited pyroclastic deposits are particularly critical here, as these, in contrast to more compact lava outpourings, can hardly form a stable permanent abutment against the pressure of the dammed water. The heat released during a volcanic eruption melts snow and ice on high volcanic flanks, creating sediment-rich mudflows called lahars. These debris-flow-like streams can deposit thick sediment layers in a short time and might block tributary valleys. After an eruption, the emptied underground magma chamber may collapse, resulting in a caldera at the surface. Such a caldera can subsequently fill with water, whereby it is a question of the stability of the slopes of the volcano whether and for how long they can withstand the water pressure.

Wind-displaced **dunes** are also able to backwater flowing waters. At first, this may seem surprising, since fine sands are extremely permeable and hardly seem to retain water. On the other hand, one has to look at extensive dune fields in desert areas to realise that even this permeability is naturally lower than that of an open channel. Thus, it is only a question of the extent of the sandy areas and the amount of inflowing water that this can be impounded. Svendsen et al. (2003) evaluated numerous studies on this phenomenon in the context of their investigations in Namibia. They interpret deposits in the southern dune field of the Skeleton Coast as traces of density currents which, starting from episodically flowing rivers, are capable of breaking through the dune walls and displacing large boulders in the process. Slack-water deposits in the adjacent Namib Desert are also interpreted as evidence of a formerly wetter climate with significant floods (Heine 2004). A special case of dune dams is reported from Antarctica (Gore 1992), where drift snow developed into an ice dam. This temporarily dammed up meltwater runoff before it yielded to pressure, causing unexpectedly large masses of water to pour through the otherwise distinctly dry Vestfold Hills in East Antarctica.

Among **organic dams**, the elaborate and extensive structures of beavers are probably the best known. In North America, for example, the number of beaver dams has declined significantly since European settlement, but the water masses released when one fails cause considerable damage (Butler and Malanson 2005). Typically, beaver dams fail during floods because they cannot withstand the increased water volumes. As a consequence, the flood wave is again significantly increased by the released water, up to amounts of 3–4 m being observed.

Artificial dams and barrages as well as barrages along river courses are numerous. They serve flood protection, water supply, energy generation and irrigation as well as tourism. The first dams were probably built in Egypt around 5000 years ago (Smith 1971; Violett 2007). Unfortunately, not all dams were able to withstand the backwatered water masses, which led to the destruction of dams and devastating floods as a result of incorrect construction methods, but also warlike attacks.

4.2 Drainage and Outburst Mechanisms

The different types of dams have different weak points that can lead to failure and the release of the dammed water masses. Basically, the stability of a dam depends on the

Table 4.1 Comparison of the characteristics of recent and historical ice-, moraine- and landslide-dammed lakes (modified including changing of data after Korup and Tweed 2007)

	Ice-dammed lakes	Moraine-dammed lakes	Landslide-dammed lakes
Catchment position	Mainly in headwaters	Mainly in headwaters	Unspecific, depends on slope stability
Cause of formation	Glacier advance; sub-, en- and supraglacial ponding	Negative mass balance results in glacier retreat	Climatic or seismic trigger, undercut slopes
Mode of formation	Quasi-periodic to episodic; gradual	Gradual	Stochastic; rapid
Dam height	<200 m	<400 m	<1100 m
Dam length	Not specific	Not specific	<6000 m
Dam width	Not specific	Not specific	<17,000 m
Dam material	Ice	Poorly sorted, massive to bouldery material, ice-cores possible, permafrost possible	Poorly sorted, typically angular
Dam volume	Not specific	Not specific	10^3–10^{10} m^3
Lake volume	<10^9 m^3	10^4–10^7 m^3	10^6–10^{10} m^3
Triggers for outburst	Floatation; enlargement of meltwater channels and tubes in and below the ice; high temperature and rainfall	Displacement waves; rainfall; increased meltwater input; ice-core and permafrost meltout; seepage; spillway incision; erosion	Overtopping; seepage; displacement waves; spillway incision; erosion; gravity collapse
Flood volume	10^4–10^9 m^3	10^4–10^7 m^3	10^5–10^9 m^3
Flood discharge	10^2–10^6 m^3/s	10^3–10^4 m^3/s	10^2–10^5 m^3/s
Life span	Not specific	10–10,000 years	0.01–10,000 years

material composition, its depositional conditions, and its geometry, which allow processes to take effect with varying intensity that can ultimately lead to lake outburst floods. Table 4.1 lists the main characteristics of lake development, size and outburst flood using recent ice, moraine and landslide dams as categories.

Ice reservoirs draw their water either from the meltwater of the glacier itself or from surface inflows from the surrounding area. They form quasi-periodically to episodically through repeated filling and draining, with the lake growing gradually according to the inflow. As the ice of the dam is in motion, the drainage pathways can quickly close again and the lake can re-form after an outburst. Ice dam heights of up to 200 m and lake volumes of up to 1 billion cubic metres have been observed nowadays, although no clear dam lengths and widths can be given, for example, in the case of damming by ice sheets (Fig. 4.2). Specific triggers of lake outburst floods are the floatation of the ice dam at sufficient water depth due to its lighter weight compared to the water (Tweed 2000). Drainage paths on, in and under the ice are steadily expanded by the flowing meltwater (Fig. 4.4), unless the movement and weight of the ice keep this in check by plastic deformation (Fountain and Walder 1998; Tweed and Russell 1999; Raymond and Nolan 2000).

Well-developed drainage pathways in the ice can be abruptly blocked again by post-fracture of the partitions and lead to a backwater of large amounts of water in the fragile network of drainage tubes. Mechanical fractures are also possible due to the pressure of the water backed up in this way. Of course, ice is particularly susceptible to weathering, so that the number of outburst events increases during warm weather and high precipitation.

Moraine reservoirs form when a glacier retreats (Clague and Evans 2000), with both the tongue-shaped basin filling with water and the surrounding ring of lateral and terminal moraines acting as a dam (Fig. 4.5). As with ice reservoirs, not the entire body of the moraine acts as a dam, so no specific dimensions can be given for this. After the active ice margin has receded, walls of terminal and lateral moraines remain, which dam up precipitation and meltwater from the glacier. Filling of the basin eventually leads to overflow, which can deepen further due to retrogressive erosion and might culminate in a lake outburst. The thawing of dead ice or permafrost in the moraines can also lead to a dam failure.

A gradual development of a moraine-dammed lake occurs, as the lake grows according to the inflow, but at the same time is drained by percolation through the dam. The moraine, which is characterised as poorly sorted due to its broad grain size spectrum, may contain ice remnants from the thawing glacier and remain preserved under the protective sediment cover for a comparatively long time. If melting finally does occur, cavities form into which the surrounding sediments sink, thus lowering the moraine wall. Typically, moraines are initially stabilised by permafrost, because low temperatures prevail at high altitudes, which means that the subsoil can remain frozen all year round. With warming, this stabilising effect of the permafrost begins to weaken and the sometimes high and steep moraine embankments run the risk of sliding and flattening

Fig. 4.4 Subglacial drainage channel widened by melting (*Photo* Daniel Garcia-Castellanos)

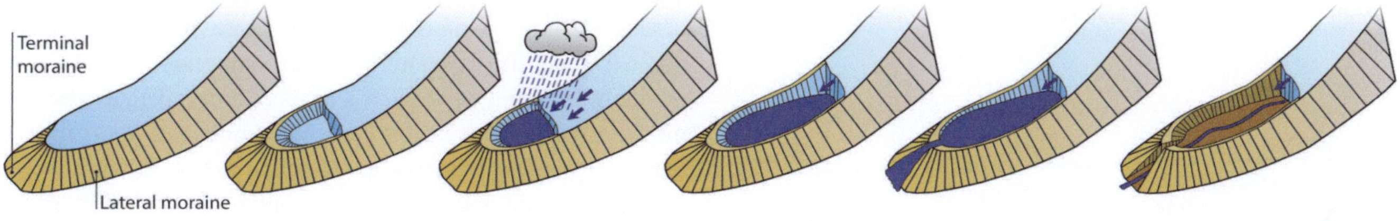

Fig. 4.5 Development of a moraine-dammed lake

due to gravity. The percolation through the dam can enlarge by piping erosion in the dam or at the source outlet, which erosively expand to such an extent that a lake outburst occurs (Fig. 4.6).

Another major trigger of outburst floods is increased inflow, whether due to precipitation or meltwater runoff from the glacier above. Emptying of water pockets in the glacier can, for example, lead to a cascade of flood waves if, after the ice-dammed lake drains, the subsequent moraine dam also fails. Furthermore, impulse waves caused by rock masses falling into the lake or glaciers calving (Vischer and Hager 1998) are important, as an overtopping wave can strongly erode the moraine flanks. The age of moraine reservoirs tends to be limited, as the moraines cannot continuously withstand water pressure. However, after an outburst, residual lakes may remain in the tongue basin until it is silted up by glacier-derived glaciofluvial sediments (Fig. 4.3).

Landslide reservoirs develop unspecifically along a river course according to the location of a landslide. The trigger for the mass movement is typically a soaking and thus weight increase of the rock masses on the slope, a slope undercutting by a river on the valley floor or an earthquake that completely loosens the rock masses. The displaced rock masses can form a dam several kilometres long and wide. In its upper reaches, depending on the inflow, a rapidly growing reservoir is formed, which can reach a very large size. Characteristic triggers of a lake outburst are similar to those of moraine reservoirs, although overflowing is observed more frequently due to the greater mass. The spillway can incise significantly due to retrograde erosion that a gap sufficient for an abrupt drainage by an outburst flood is generated. (Fig. 4.7).

The loosely deposited rock masses have cavities through which water can seep in large quantities. This creates several partly independent, partly interconnected drainage systems. These have springs in different positions (Fig. 4.6), which incise prominent channels into the dam and thus destabilise it by erosion (Fig. 4.8).

While the lower lake is drained by a spillway, the upper lake has no surface drainage (Glazirin 2013). A currently dry channel bed is clearly visible on the landslide mass.

Fig. 4.6 Spring of seepage water at the dam of Upper Ikhnach Lake, Uzbekistan (cf. Fig. 4.1). Depending on the water level in the reservoir, different drainage pathways are active, whose outlets in this case lie high above the dam base and from there begin to incise (person for scale)

Fig. 4.7 Spillway of Lower Ikhnach Lake as seen from below. The course of the stream has already undercut the slopes as can be seen from the interrupted vegetation cover on both banks

This feeds water on the surface to the lower lake from a spring pool within the landslide mass when the water level is high during the snowmelt in spring. The indication of a potential spillway at the proximal part of the dam is relict, as indicated by the extent of the fresh rockfall deposit at the base of the slope, which extends into the notch.

Experience shows that a landslide lake has either a very short or a very long life span. This depends on the amount of displaced rock on the one hand and the runoff in the blocked valley course on the other. The elevation of small amounts of debris is quickly reached and subsequently removed if runoff is sufficient, whereas voluminous and extensive deposits can form a long-lived dam in which a stable overflow develops. In an empirical study, Costa and Schuster (1988) document that of 73 landslide lakes, 50% had a life span of one week, 85% of less than one year (Fig. 4.9).

The above considerations have been differentiated according to the origin and type of dam, which usually requires detailed on-site investigations and expertise to identify the relevant structures and components. Manville (2001) developed a pragmatic approach in terms of focusing solely on the resulting outburst flood. Since the processes leading to dam failure are more important in this context than the formation of the dam itself, he focuses on their frequency. Thus, it can be observed that overtopping is the most common cause of a outburst flood from a natural reservoir, namely in landslide lakes. As previously indicated, a distinction must be made as to whether the overflow is caused by a continuous rise in the lake level or by an impulse wave. Using an example from the Argentine Andes, Hermanns et al. (2004) show how a landslide dam that had been stable for thousands of years was unexpectedly overflowed and destroyed by an impulse wave. The resulting outburst flood in turn led to the failure of other natural reservoirs below, triggering a cascade of events.

In historical times, the Goldau Landslide, which fell on the Swiss village of the same name in 1806, gained sad notoriety (Bussmann and Anselmetti 2010). Not only was almost the entire village and its inhabitants killed, but an impulse wave was also formed in the neighbouring Lauerzer Lake. The wave reached a height of around 15 m and cost the lives of another ten people who were near the shore.

Following in the frequency of triggers of dam failure are slides below the water level into the lake. This is due to the different natural slope angle of sediments in dry or wet conditions. If rocks have previously slid below the water level, the sediments above lack the abutment, so that they also slide down, reducing the overall height of the dam (De Groot et al. 1995).

Fig. 4.8 Lower and Upper Ikhnach Lake from above the starting point of the landslide, whose deposits form the dam of the upper lake (cf. Fig. 4.1)

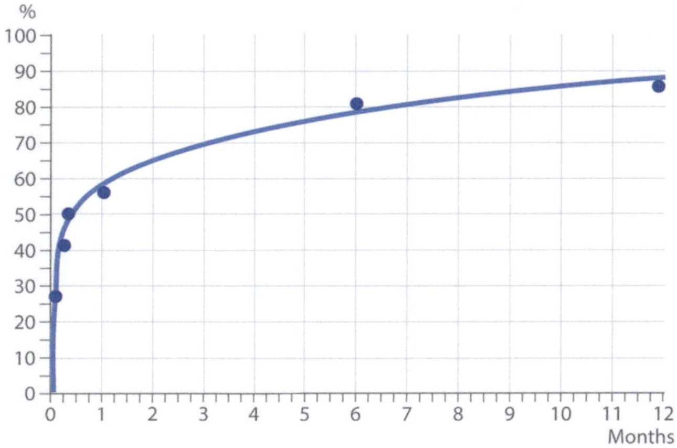

Fig. 4.9 Age of landslide dams until dam failure (modified after Costa and Schuster 1988)

Finally, the flow through the dam is worth a closer look. In this process, dissolution processes in the sense of chemical weathering of the dam components can take place. The aforementioned widening of the paths due to melting ice, the piping erosion and erosion at spring outlets are additional component in this context. Here it becomes clear that in practice a distinct separation between the sub-processes is not unambiguously possible and usually several processes take place in parallel or one after the other. This differentiation of the processes is important with regard to the dynamics of an outburst flood. Generally speaking (cf. Tweed and Russell 1999), it can be summarised that abrupt failure of a dam results in an extremely steep rise in the flood hydrograph, while gradual widening results in a correspondingly slower rise in flood discharge. Dam failures with gradual widening also last longer (Fig. 4.10).

These general patterns of outburst flood wave hydrographs do not fundamentally change downstream, but they flatten out due to backwater in the inundated area. As a result, the peak discharge decreases, but the duration of the increased water level increases. From the conceptional point of view, these different pattern is logical and plausible, while the empirical evidence is less clear (Herget et al. 2015).

In unfavourable topographical conditions, for example due to a bottleneck below the dam, the initially paradoxical phenomenon of an increase in the downstream discharge can also occur. This is possible if the advance speed of the front of the outburst flood wave is reduced by the bottleneck, so that the water flowing downstream accumulates. If one then compares the discharge hydrographs with each other, a higher peak discharge may have developed below the bottleneck than immediately below the dam.

The effect of earthquakes on reservoirs is also remarkable. One would generally expect earthquakes to endanger the stability of dams. Accordingly, earthquakes are listed among the triggers of outburst floods in numerous publications. On the other hand, Hermanns et al. (2004) document for the earthquake-rich southern Andes that dams can survive for millennia despite repeated severe earthquakes. Similarly, Post (1967) lists the magnitude 9.2 earthquake in Alaska in March 1964, where none of the more than 700 ice-dammed lakes under observation drained. Undoubtedly, there are some natural reservoirs whose dams have failed due to earthquakes, as indicated for the June 1786 event in China (Dai et al. 2005). As a result of a magnitude 7.7 earthquake, a landslide occurred that quickly dammed the Dadu River into a 70 m deep lake with a volume of 50 million m^3. After only 10 days, the dam broke, releasing an outburst flood with an estimated peak discharge of

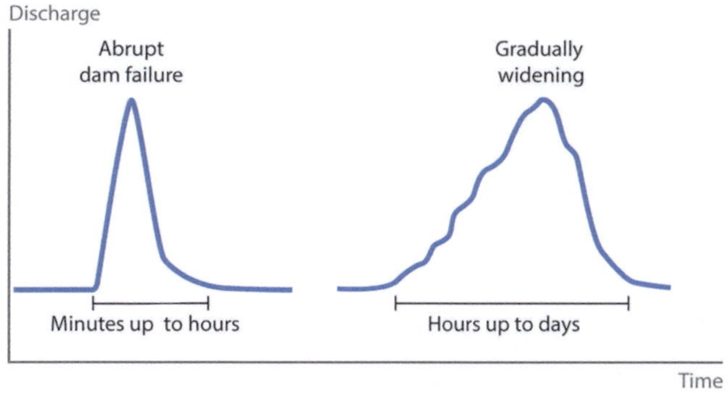

Fig. 4.10 Schematic hydrographs of lake outburst floods with abrupt dam failure and gradual widening of the outlet (modified after Costa 1988)

37,000 m^3/s that probably claimed the lives of about 100,000 people. An aftershock is cited as the trigger. Hydraulic analyses of the dam breach and the flood wave, on the other hand, have shown that the outburst could only have been caused by an overflow (Chanson 2005). The essential argument is that, physically speaking, a dam of the dimensions that have been handed down cannot open completely abruptly simply because of its size. Taking into account the alternative failure mechanism, the peak discharge is still around 6000 m^3/s, which roughly corresponds to the maximum measured flood discharge. Certainly, it cannot be ruled out that aftershocks had a negative influence on the bedding conditions of the landslide mass, but they were not a direct triggering process of the outburst flood. Here it is important to carefully distinguish between cause, effect and subsequent processes of the actual dam failure. In this context, reference is again made to the example from the Argentine Andes by Hermanns et al. (2004), who identified an impulse wave as the cause of the outburst after otherwise obvious stability. The impulse wave, in turn, was caused by rock avalanches falling into the reservoir as a result of an earthquake. To those who think this is sophistry—the trigger is still an earthquake, only that another intermediate step has been identified—it should be pointed out that rock avalanches can also be triggered without an earthquake. This puts into perspective the general statement that earthquakes are a major cause for the failure of natural dams. By the way, impulse waves caused by falling rocks reached heights of 524 m in the Lituya Bay in Alaska in July 1958, with sufficient energy to cut metre-thick trees to a height of over 200 m (Miller 1960; Fritz et al. 2001). As a result of rockfalls into the bay, it has repeatedly happened that such impulse waves like tsunamis with advance speeds of estimated 120 km/h have removed the forest along with the ground cover.

Summarising the considerations of the different dam types and breakout mechanisms, Fig. 4.11 provides an overview of the magnitudes of reservoir outburst floods in historical and modern times. Different types of dams and failure mechanisms are taken into account.

The figure illustrates that, as expected, there is a correlation between the amount of water discharged and the peak discharge. This correlation is the background for an estimation procedure of the dimension of lake outbursts (cf. Sect. 2.4). On the other hand, some outliers stand out, indicating that special conditions prevailed during the development of the outburst flood. An example of such specific conditions is the manmade reservoir drained by spillover with a discharge of about 100 m^3/s and a lake volume of about 100 m^3 (Fig. 4.11). In this case study, another outburst flood wave flowed through a small retention basin and destroyed its dam without any delays.

The modelling of reservoir outbursts is faced with the problem of, on the one hand, wanting consideration of sub-processes on a physically based basis for the realistic estimation of the discharge hydrograph of a lake outburst, and, on the other hand, being able to quantify the corresponding influencing factors in sufficient detail. Expecting a favourable data situation, models are able to predict the peak discharge of a dam break with 30% accuracy, while for other parameters the accuracy is lower (Morris et al. 2008). In order to be able to reliably estimate the stability of natural and artificial dams, numerous modelling approaches are available from the engineering sciences (Singh 1996; ICOLD 1998; Wahl 1998; Morris et al. 2008; Wu 2011; Westoby et al. 2014). These can only be applied to natural dams with their rather irregular shapes and unsystematic composition to a limited extent due to the parameterisation referring to constructed dams. Physically based modelling approaches are also available for estimating the development of drainage pathways in and on a glacier (e.g. Raymond and Nolan 2000; Clarke 2003; Worni et al. 2014).

Within the framework of basic research, numerous detailed processes that are of central importance for explaining the processes of dam failures have been investigated. In order to be able to give a physically based explanation, this requires numerous parameters such as water temperature, grain size distribution in a terminal moraine or geometry of a drainage tube in a glacier, which can hardly be determined, certainly not for the reconstruction of previous events. It must also be taken into account that many natural lakes for which a hazard assessment is to be carried out are located in remote high mountain regions that can only be reached with difficulty and for a short time in high summer. Here, one can fall back on data from remote sensing through satellite images, which of course cannot then provide the mentioned information for physically based models (cf. e.g. Huggel et al. 2002).

Ultimately, one still has to rely on the comparatively simple empirical model approaches, which can be verified well and clearly by simulations in hydraulic test channels in a process-oriented manner (Carrivick et al. 2011). However, a plausibility test of the results remains indispensable. It is important here to always be aware of the admissibility of the application space or aspect, because a comparison of the estimates from different approaches delivers extremely different values, as expected. Manville (2001) proves this with peak discharge estimates that vary between 65 m^3/s and 14,106 m^3/s for an identical model reservoir outburst if genesis and outburst mechanism are disregarded. Wahl (2004) shows the same for a hypothetical breach of an artificial reservoir whose peak discharge varies between 2210 and 72,200 m^3/s when regression equations are applied without reflection. Anyhow, the arguments for empirical models remain, since the parameters required for detailed models cannot be recorded representatively with acceptable effort or at all. This is especially true for the reconstruction of past events, in which no dam may be preserved.

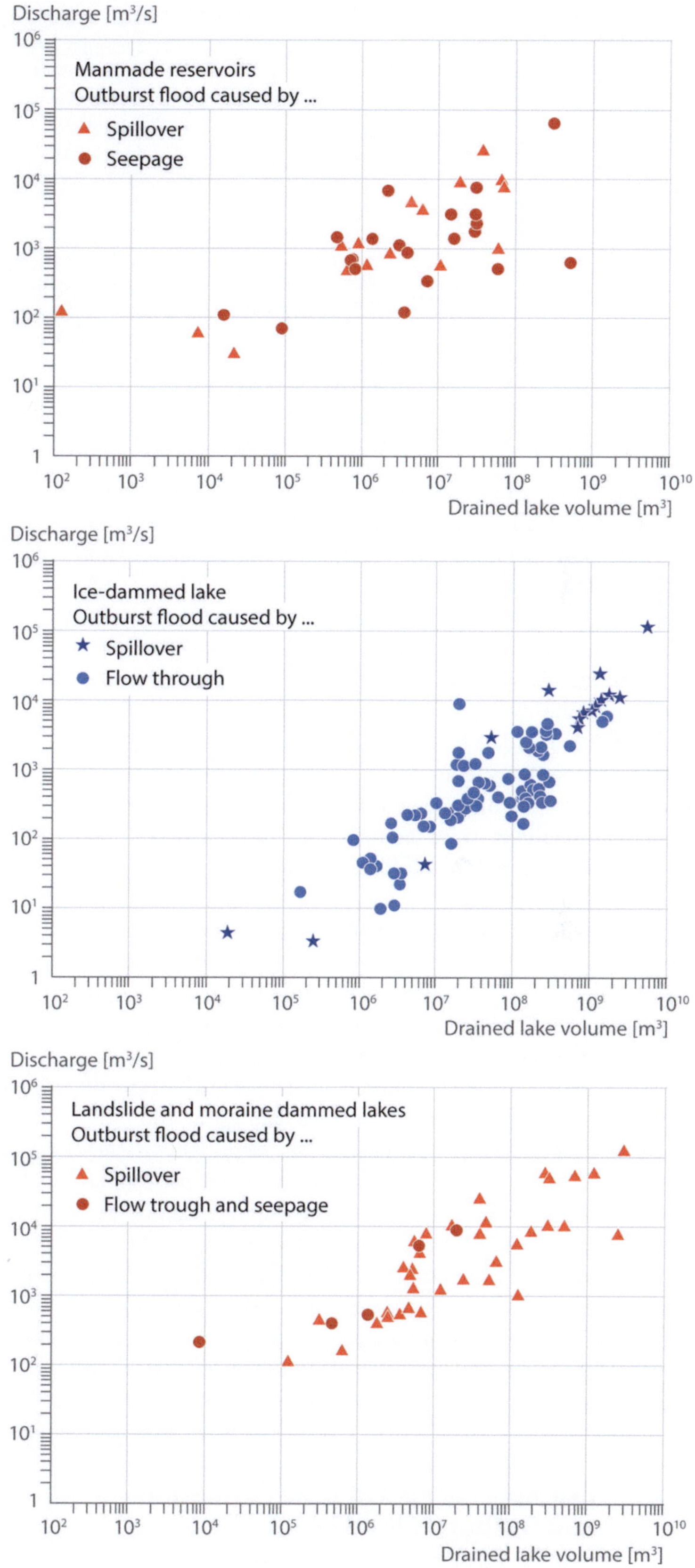

Fig. 4.11 Peak discharge in relation to drained lake volume of recent and historical lake and reservoir outburst floods (data from Walder and Costa 1996; O'Connor and Beebee 2009)

Table 4.2 Comparison of regression equations for estimating the peak discharge Q (m^3/s) of lake outburst floods (Costa 1988)

Type of lake	Dam height H (m)	Lake volume V (10^6 m^3)	Dam factor (H·V)
Artificial lake (n=31) 29.5<Q<65,136	**$Q=10.5\ H^{1.87}$** 1.8<H<83.8 $r^2=0.65$ Sd=82%	**$Q=961\ V^{0.48}$** 0.0038<V<703 $r^2=0.65$ Sd=124%	**$Q=325\ (HV)^{0.42}$** 0.0066<HV<26,020 $r^2=0.75$ Sd=95%
Landslide lake (n=10) 64<Q<56,650	**$Q=6.3\ H^{1.59}$** 10.7<H<274 $r^2=0.74$ Sd=147%	**$Q=672\ V^{0.56}$** 0.105<V<670 $r^2=0.73$ Sd=142%	**$Q=181\ (HV)^{0.43}$** 1.12<HV<128,000 $r^2=0.76$ Sd=129%
Ice-dammed lake (n=21) 15<Q<22,650	**$Q=21.6\ H^{0.73}$** 35<H<620 $r^2=0.08$ Sd=236%	**$Q=113\ V^{0.64}$** 0.135<V<1730 $r^2=0.80$ Sd=106%	**$Q=3.8\ (HV)^{0.61}$** 576<HV<802,500 $r^2=0.79$ Sd=75%

r^2—correlation coefficient
Sd—Standard deviation
n—number of data

Costa (1988) has developed corresponding regressions for ice-, landslide-dammed lakes and artificial reservoirs and can show that the standard deviation and the correlation measure generally improve when more than one input variable is used. Specifically, he illustrates this with comparisons of the input variables dam height H and lake volume V as well as a so-called dam factor, which results from the product of the two values (Table 4.2). For the case studies he considered in developing the regression equations, Costa also developed envelopes for expected maximum peak discharges. Focusing on an engineering perspective, additional empirical relations were compiled by Wu et al. (2011), while Westoby et al. (2014) focus on the same topic from a geoscientific perspective.

A next higher level of resolution with regard to process differentiation is formed by regression equations, in the development of which the respective rupture processes have been taken into account by the formation of subgroups. For example, Walder and Costa (1996) distinguish for ice dam outbursts between drainage through tunnels in the ice ($Q=46\ V^{0.66}$, $r^2=0.70$) and breakage of the ice dam ($Q=1100\ V^{0.44}$, $r^2=0.58$). For naturally dammed lakes, especially if they have just recently formed and are still growing, the estimation of the lake volume is challenging. Surveys can be carried out by combinations of echo sounder and GPS in a rubber boat (Fig. 4.12) but logistical problems to access the lake remain. Consequently, tools to estimate the lakes volume by its surface—which can be determined by remote sensing approaches—were developed (Cook and Quincey 2015) and the correlations are better than one might assume.

Fig. 4.12 Bathymetric survey based on GPS and echo sounder on a glacial lake in Tien-Shan Mountains, Kazakhstan

4.3 Examples of Recent Lake Formation and Outburst Flood Events

Using selected examples, a spectrum of lake outburst floods resulting from dam failures is presented below. This is not a systematic, encyclopaedic overview, but rather a presentation of the variety of phenomena and magnitudes, which is supplemented in the following chapter by the megaflood events during the last ice age. In some of the lakes presented, there are clear indications of increasing instability of the dam, and their outburst is still imminent in the near future.

4.3.1 Ice-Dammed Lakes

Among the numerous recent ice-dammed lakes in the Alps (e.g. Haeberli 1983), the Märjelensee on the Great Aletsch Glacier in Valais is probably the best known, along with the lakes in the Austrian Ötztal (Patzelt 1994; Krainer and Spieler 1999; Nicolussi 2013). Here, the Great Aletsch Glacier dammed the runoff originating from a tributary

Fig. 4.13 Lake Märjelen on the Great Aletsch Glacier in September 1909 (Lütschg 1915). At the time of the photograph, the water level had lowered somewhat shortly before, as shown by the stranded ice blocks. The change in vegetation cover on both sides of the valley indicates the maximum level of the lake, whose level was about 30 m higher in July 1878

valley to form a lake that was episodically drained through the glacier (Fig. 4.13) (Lütschg 1915).

Lake Märjelen repeatedly dammed up to 77 m deep, so that local watershed devides were reached and the lake found an overflow into the adjacent Fieschertal, where this additional constant water supply led to flooding. Particularly critical, however, were the repeated draining of the lake through the Aletsch Glacier, which still made itself felt in Brig, barely 20 km away, with a 2 m high flood wave. This drainage was possible despite the maximum level of the Great Aletsch Glacier during the so-called Little Ice Age (Holzhauser 1997). In addition to the destructive force of the flood itself, great damage was caused over a wide area on the valley floors, especially by the covering with sands and gravels. The flood waves were noticeable along the entire Rhone valley up to the mouth of Lake Geneva. In order to avoid this damage, measures were taken early on so that less water could accumulate in the lake. First a ditch, and later a deeper gully, was dug into the watershed to the neighbouring valley to force an overflow even at lower water levels. Later, this was reinforced by an artificial reservoir, whose water is safely discharged in a pipeline through an additional tunnel, which largely interrupts the inflow into Lake Märjelen (Fig. 4.14). Through this measure, no significant amount of water can accumulate and Lake Märjelen has essentially become history.

The Unteraar Glacier, source of the Aare River in Switzerland and located near the Grimsel Pass, is a source of particularly remarkable reservoir emptying. Here, bulges and increased flow velocities were observed early on, which were associated with the emptying of englacial water pockets (Iken et al. 1983). These meltwater releases could be safely buffered in the artificial reservoir created directly below the glacier to generate electricity. During such outbursts, wood remains of stone pines, pine trees, larches and other tree species as well as compressed peat remains were flushed out at the Unteraar Glacier. These findings prove that there must have been times when the glacier was much smaller. Trees grew and bogs existed in the areas now covered by the glacier, which can be interpreted as an indicator of a much warmer climate (Hormes et al. 1998). These finds, which could be dated by radiocarbon and dendrochronology, are not scattered in terms of age over the entire Holocene, but form distinctive time periods that can also be confirmed on other glaciers and should therefore be regarded as a regional phenomenon for the entire Alps (Hormes et al. 2001; Ivy-Ochs et al. 2009a, b). Similar wood and peat finds, for example from the Steinlimi glacier (Fig. 4.3), prove that here, too, large areas that are now covered with alpine mats were covered with forest in Roman times (Schlüchter and Jörin 2004). At that time, the glacier tongues ended at least 300 m higher than today, so that they were hardly visible when crossing the passes and were not perceived as an obstacle. The maximum extension of the glaciers during the Little Ice Age reached at about AD 1850 was obviously the exception and not the rule since the end of the last glaciation about 10,000 years ago. This also puts the well-documented shrinking of Alpine glaciers in historical times into perspective, even if the consequences of global warming and its anthropogenic influence are by no means mitigated.

The ice-dammed lake outburst floods on **Iceland** have become eponymous for the phenomenon with the term “Jökulhlaup” (glacial flood). In addition to the lakes dammed by stream blockage by glaciers, such as the Graenalon on Skeidararjökull presented at the beginning (Fig. 4.2), it is the outburst flood waves triggered by subglacial volcanism that are of particular interest here (Björnsson 2002, 2009, 2016). In recent decades, subglacial eruptions have occurred regularly, almost periodically, in the Grimsvötn caldera under Vatnajökull, for example. These eruptions, which occur on average every 4.6 years, generate outburst floods that last 2–3 weeks and typically have discharges of 600–10,000 m^3/s. Particularly noteworthy is the interplay of forces between the heat source under the ice, the water formation immediately above it and the surcharge by the ice sheet, which is about 150–270 m thick above the caldera. As a rule, even during a subglacial eruption, the overlying ice sheet remains intact despite subsidence of the surface. Consequently, the resulting meltwater is under pressure. Until an outburst occurs, the water has to travel about 50 km under the ice, whereby the gradient, the local topography and subglacial meltwater paths, as well as the glacier flowing much easier and faster on the water, determine the timing in a complex interplay of forces (Roberts 2005; Russell et al. 2007; Björnsson 2010, 2016).

With a peak discharge of about 50,000 m^3/s, the eruption flood wave from Vatnajökull in November 1996 was

Fig. 4.14 Artificial reservoir above Lake Märjelen. On the opposite flank of the valley, one can still see the horizontal change in vegetation cover that marks the maximum water level of Lake Märjelen. The mouth of the drainage tunnel at the access road is also visible

exceptionally large (Russell et al. 2006; Björnsson 2002). The eruption started from a newly formed subglacial volcano in late September 1996 and was strong enough to penetrate the 500–700 m thick glacier there. It produced about 3,800,000,000 m^3 of meltwater, which drained subglacially and initially filled the Grimsvötn caldera under the ice. Finally, a month later, in the morning hours of November 5th, the outburst began, which, in accordance with the subglacial catchments, took place as usual to the south over the Skeidarar Glacier, a tongue of the Vatnajökull ice shield. After preannouncement by vibrations from the ice body, the peak flow developed within only 14 h, fed by numerous outlets on the glacier tongue and partly newly formed glacier gates over a width of 23 km. In numerous places the glacier front burst and icebergs the size of houses were carried onto the outwash plain (Fig. 4.15). The sediments deposited downstream caused a temporary shift of the coast by about 900 m seaward, which was subsequently eroded by the waves of the North Atlantic.

Additionally in prehistoric times, large outburst floods were initiated by the subglacial volcanoes under Vatnajökull, which took their way northwards across the Jökulsa a Fjöllum valley. Here, they have eroded large and deeply incised valleys, although the peak discharges responsible for this are still being assessed differently: Carrivick et al. (2004) found evidence of a debris flow and estimated peak discharges of almost 50,000 m^3/s. Alho et al. (2005) modelled values in the order of 900,000 m^3/s for peak discharge assuming regular flow conditions.

The largest outburst flood in historical times in Iceland occurred on 12th of October 1918 (Tomasson 1996). An eruption of Katla Volcano, covered by Myrdalsjökull in southern Iceland, began with a significant earthquake, which was immediately followed by an unmistakable rush of flowing water and rumbling of entrained boulders. The volcanic eruption melted large quantities of ice by the hot ash and outflowing lava, which immediately began to tumble down the volcano's flanks. The total volume of water is estimated at 8 km^3, with water temporarily stored by channel blockages caused by collapsed channel walls and icebergs, resulting in a multiphase outflow. According to eyewitness reports, an initial wave of meltwater, ice blocks and entrained rock had an advance speed of around 10 m/s. The deposits of the outburst currents, which took different pathways, can be quantified from historical records despite subsequent floods and resulting coverage. Due to the different paths and several phases, no representative discharge value can be given for the event as a whole. However, flow velocities of 100 km/h and discharges of around 300,000 m^3/s were achieved locally.

Outburst floods of similar magnitude are also reached recently at the Hubbard Glacier in Alaska. The glacier flowing into the "Disappointment Bay" near the village of Yakutat (an early explorer hoped in vain to find access to the Northwest Passage here) repeatedly advances at increased flow rates and then blocks Russell Fjord, which flows into the bay (Fig. 4.16). As a result, precipitation and inflowing meltwater accumulate in the fjord, and Russell Fjord becomes Russell Lake.

Fig. 4.15 Outburst flood wave of the subglacial ice reservoir at Skeidarar Glacier, Iceland, November 1996

Of the last repeated periodic advances of the glacier, those of 1986 and 2002 led to a blockage of the fjord, a build-up of water and finally an outburst flood (Mayo 1989; Trabant et al. 2003). The next damming is expected for the near future even though the delicate balance of precipitation, temperature and meltwater dynamics within the glacier are difficult to predict (Stearns et al. 2015). The impoundment in May 1986 reached a water height of 25.5 m and a volume of 5.4 km^3. At a height of 39 m, overflows into the Situk River would start to function. This happened repeatedly in prehistoric times, as can be inferred from the over-wide channel bed of the river already in its upper reaches. The outburst flood wave that finally poured into Disenchantment Bay in October 1986 had a peak discharge of about 105,000 m^3/s. Despite the impressive volume of the outburst flood wave, it was rather the development within the fjord during the build-up that proved to be problematic. Due to the interruption towards the open sea and the constant inflow of meltwater, the fjord sweetens and a stratification of light freshwater on heavy saltwater forms, which leads to a serious disturbance of the ecosystem. Ultimately, this also has economic consequences, as the fishing industry is the most important source of income for the local population. Accordingly, the closing of the fjord in June 2002 was carefully observed. Webcams were installed and the development of the ice dam could be observed worldwide. The advancing glacier pushed up an upsetting moraine of sediments from the seabed, which together with the ice masses closed off the fjord (Fig. 4.17).

By 14th of August 2002, the lake had reached a damming height of 15 m before the dam broke around midnight. From the lowering of the water level in the fjord it can be deduced that the peak discharge of the outburst flood wave was about 52,000 m^3/s. Among recent ice dam outbursts, those from the Russell Fjord thus occupy a top position in terms of peak discharge.

One of the numerous subglacial lakes in Antarctica is Lake Vostok, not far from the Russian research station of the same name (Siegert 2005; Siegert et al. 2005). Among the various suggestions for the formation of the 50 × 200 km large lake with a volume of about 5400 km^3, one is that the lake was formed before the beginning of the glaciation of Antarctica. This would give the lake an age of several million years. It is remarkable that the lake water exists completely isolated from the global water cycle. Due to this long isolation, it is assumed that microorganisms exist in the lake that have followed a completely independent path in evolution. For this reason, people have so far been reluctant to sample the lake itself in order to avoid contamination with the atmosphere. Even if the lake no longer contains water from the time of its genesis before the onset of glaciation, the time period is sufficient for the possible evolution of unique organisms in the lake. The nearly 150 subglacial lakes under the Antarctic ice sheet are preserved by the geothermal heat flow from the Earth's interior, which is protected from the cold atmosphere by the ice sheet, similar to an igloo. In addition, the pressure melting of the ice in the subsurface is caused by the overloading of the ice sheet, which is around 3700 m thick at Lake Vostok. Against this background of pressure melting, Erlingsson (2006) considers the stability of the subglacial water basin, especially since there is evidence of water exchange

Fig. 4.16 Yakutat Bay, southeastern Alaska, with the advancing Hubbard Glacier and later Russell Fjord dammed into a lake (USGS satellite image from Molnia 2008)

between the lakes (Wingham et al. 2006). They hypothesise that after millions of years of quiescence, the pressurised lake water could one day flow out, initiating a jökulhlaup in Antarctica. Due to the size and extent of the ice cover, this would be lost in the vastness of Antarctica despite the large volume of sea water and thus hardly lead to a remarkable outburst flood. On the other hand, it is to be expected that a new ice stream of increased flow velocity could be generated by the reduced friction of the ice sheet on the bed.

From the abundance of examples of ice dams and their outburst floods (Walder and Costa 1996; Tweed and Russell 1999; Worni et al. 2014; ICIMOD 2023; Veh and Lützow 2024; NVE 2024), only a small selection with various special features could be presented above. It is particularly noteworthy that ice dams can close again after an outburst flood due to the movements of the glacier, dam up and drain again, while the other dams remain out of function after the drainage of the lake as the gap does not close again.

4.3.2 Lakes Dammed by Landslide Deposits

The 1991 landslide in the valley of the Matter Vispa near Randa in the Swiss Alps received a lot of attention as the access to the famous village of Zermatt was interrupted (Schindler et al. 1993). Here, a total of 33 million m^3 of

Fig. 4.17 Advance of the Hubbard Glacier in 2002, formation of the ice reservoir in Russell Fjord and final phase of the outburst flood (photos by the US Geological Survey and US National Park Service)

debris fell downhill in phases beginning on the morning of 18th of April, which led to an interruption of the connection to Zermatt and created a reservoir. Through the use of pumps and the construction of a drainage channel, it was possible to largely prevent the lake from rising to a level that would endanger the stability of the dam, so that an outbreak of the reservoir could be prevented despite inundations in the village (NGO 1991).

The failure of a landslide dam in the Partnachtal valley in the Bavarian Alps in August 2005 was not quite so fortunate (Morche et al. 2007). After long-lasting precipitation with intensities of up to 230 mm per day, a landslide dam that had previously been stable for years gave way, so that the released water masses of the lake increased the magnitude of the flood caused by precipitiation. The lake itself was dammed to a volume of 9800 m^3 and a depth of 2.5 m by a landslide that occurred between 1789 and 1806. According to empirical formulas, a peak discharge at the outburst of 26–33 m^3/s can be expected, which may not seem much at first, but had devastating effects in the small mountain valley. The main reason is the mobilisation of 100,000 t of gravel and debris in the valley bottom in the form of debris flow surges, which led to a continuous reshaping of the valley floor.

One of the largest landslides in the world occurred shortly after the end of the last ice age in the Vorder Rhine valley upstream of Lake Constance in the Swiss canton of Graubünden (Fig. 4.18).

In the area where the Hinter Rhine flows into the Vorder Rhine, two landslides occurred in quick succession near the present-day villages of Tamins and Flims. Although discussions about the details are still ongoing, it appears that thawing permafrost at the end of the Ice Age destabilised the slopes to such an extent that the landslides with volumes of 1.3 km^3 (Tamins) and 12 km^3 (Flims) were mobilized. Deposits of small ponts formed on the surface of the landslide deposits provide a maximum age of 9480–9430 years before present (Deplazes et al. 2007), which can be reconciled with exposure ages of 8900 ± 700 years (Ivy-Ochs et al. 2009). The fact that the Tamins Landslide occurred first can be deduced from the bedding conditions (Schneider et al. 2004; Poschinger 2005). The Hinter Rhine dammed up to form Lake Bonaduz, which extended for about 20 km. The flattening of the valley floor in the lowest section of the Hinter Rhine valley, which can still be seen today, is due to its deposits, which are more than 50 m thick. The Flims Landslide detached a large part of the mountain flank over a length of about 7 km at the Flimser Stein and filled the valley of the Vorder Rhine with sediments up to 400 m thick. As a result, the Vorder Rhine dammed up to form Lake Ilanz, which is about 40 km long. By the elevation of characteristic lake deposits in the tributary valleys, it can be reconstructed that the maximum water level of Lake Ilanz reached a height of 1175 m (Wassmer et al. 2004). Other data focus on indicators in the immediate vicinity of the dam itself, which may no longer have been preserved at their original maximum elevation due to later erosion in the landslide mass. It must be taken into account at this location that Lake Ilanz was drained by spillways. These incised during the outburst drainage and subsequently left behind initially over-steepened slopes, which are still characterised by constant erosion today, as shown by the lack of vegetation cover (Fig. 4.19). The location, origin and sequence of activity of the spillways are still controversially discussed. However, the consequences of the outburst of Lake Ilanz can be found in the form of characteristic deposits at the bottom of Lake Constance (Fig. 4.18; Schneider et al. 2004). Far from the mouth of the Alpine Rhine into Lake Constance, there are two intercalations in the uniform sequence of lake sediments, which are interpreted as deposits of two outburst floods. Their age dating fits well with the lake formations in the Vorder and Hinter Rhine valleys. The sequence of events and correlations in detail (one-time entire or phased lake drainage?) are also still interpreted differently. Against this background, estimates of the peak discharge are uncertain. If the regression equations are applied (Table 4.2), values of 86,000–195,000 m^3/s result for a lake volume of 25 km^3 and a dam height of 400 m. These values only serve to estimate the magnitude of the peak discharge and must by no means be seen as the result of a site-specific modelling. It is striking that the high values are significantly influenced by the relatively large lake volume, without an abrupt release of the entire lake volume being justifiably assumed. Therefore, values in the five-digit range might be more plausible.

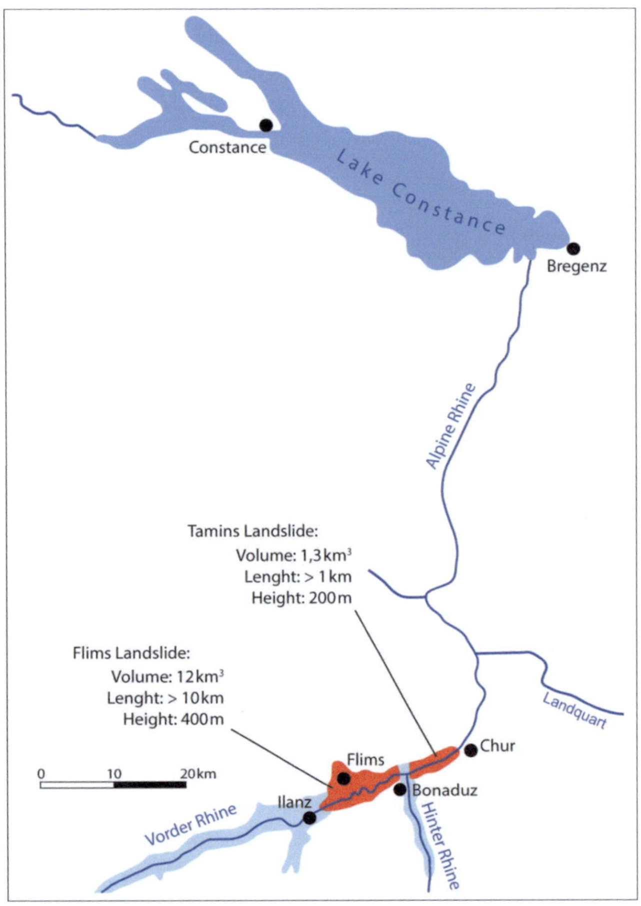

Fig. 4.18 Flims and Tamins Landslides and their lakes (modified after Schneider et al. 2004)

Hewitt (1998) describes landslide dams that had heights of more than 1100 m from the Karakoram in Pakistan. For example, in the Rondu-Mendi Basin in the Indus Valley, a multi-phase landslide occurred that cannot be precisely dated, but probably occurred before the peak of the last glaciation—orders of magnitude of 30,000 years before today are mentioned. Due to its size alone, the structure with its numerous partial spillways and structures is not easy to grasp. From deposits in the upper stream and the valley geometry, a lake length of 180 km can be deduced, but the data are too vague for reliable estimates of an outburst flood wave.

Dimensions can, however, be estimated by comparing it with the outburst flood in the Indus Valley in 1841 (Mason 1929). In the winter of 1840/41, an earthquake triggered a landslide on the flank of the Nanga Parbat Massif, blocking the Indus Valley with deposits over 300 m thick. Six months later, the dam was overtopped, resulting in a rapid breach and lake emptying within a day. Since the river below had permanently carried relatively little water due to the damming up of the lake, warnings about an imminent flood wave were not taken very seriously. The outburst flood wave with a height of more than 20 m announced itself loudly, but nevertheless thousands of people probably died. Even in Attock, 400 km away, the outburst flood wave had a strikingly steep wave front, peaking at levels 28 m above the current low water level and 13 m above the normal summer high water level. Estimates put the peak discharge at 57,000 m^3/s.

That the formation of large landslide-dammed lakes in the region is still relevant today is shown by events in spring 2010. On 4th of January, a landslide occurred in the Hunza Valley at Attabad, burying the village underneath and beginning to block the river with a dam more than 100 m high (Petley 2011). The landslide disrupted the Karakoram Highway and the rising lake level began to flood the settlements, the few cultivated areas on the valley floor, and the uphill access road to the villages (Fig. 4.20). Given the complete uncertainty of the dam's stability, especially if the water level had risen high enough for a spill, a spillway was constructed in May to provide regulated runoff before the maximum level related to maximum pressure on the dam was reached. The observed seepage through the deposits by the backwatered water was assessed as particularly critical. Flushing out of fine material within the dam is hardly noticeable, but can have a lasting effect on the stability and lead to an abrupt slipping of the mighty deposits. By the end of June 2010, an equilibrium had been established between the inflow into the lake, the water flowing through

Fig. 4.19 Deposits of the Flims Landslide as seen from the valley bottom within the Vorder Rhine valley

the landslide masses and the overflow and its incision. This can be interpreted as an indication of stability and the lake became a tourist attraction due to its beautiful appearance in the dry and barren mountain environment. Due to the successfully installed spillway the lake volume could be reduced by 72% which means about 40 m in elevation until 2015 (Delaney and Evans 2017).

A similar situation has prevailed for more than one century already in the valley of the River Murghob in the **Pamir Mountains**, Tajikistan (Alford and Schuster 2000; Risley et al. 2006). In the winter 1911, a landslide triggered by a severe earthquake filled the river valley with 500 m thick rock masses. Behind the dam named after the buried village of Usoi, the over 60 km long Lake Sarez was formed with a volume of 17 km^3 (Fig. 4.21).

Even though the lake itself is located in a remote mountain valley, it poses a particularly great danger. On the one hand, the enormous dam height plays a major role, which makes it the highest dam in the world, including all technical dams, and therefore has a high erosion potential in case of initial overflow. Furthermore, the region remains at particularly high and frequent risk from earthquakes, which can trigger further landslides into the lake with a resulting impulse wave (Stone 2009). Depending on the volume of the rock masses falling into the lake, modelling on this has shown impulse waves of 2–87 million m^3 of water sloshing over the dam crest, which could lead to peak discharges of 550–1100 m^3/s depending on the distance downstream. Such discharges may not seem particularly dramatic at first. However, if one takes into account that the river valleys in the semi-arid regions of Central Asia are the only usable living and economic areas, it becomes clear that a large part of the population along the Amu Darya River up to Lake Aral would be permanently affected by an outburst flood or damage to agricultural land. It is hard to imagine the consequences of an abrupt failure of the dam with estimated peak flows of 20,000–64,000 m^3/s. Consequently, serious concerns remain (UN-ISDR 2007). In landslide- and rockfall-dammed lakes, the lifespan until an outburst depends on the ratio of lake inflow to dam volume. However, even if such reservoirs have passed the critical juvenile phase (Fig. 4.9), impulse waves can still lead not only to an overflow wave but also to a lake outburst thousands of years later.

4.3.3 Moraine Dammed Lakes

Due to the current reduction of glaciers, numerous tongue basins worldwide with their framework of lateral and terminal moraines are becoming reservoirs. As a result, the number of reservoir outbursts is increasing, as Richardson and Reynolds (2000) can show for Central Asia, while in the meantime a reduced number of ice-dammed lakes is documented (Carrivick and Tweed 2016). Warm weather can temporarily produce more meltwater, which provides

Fig. 4.20 Landslide dam in the Hunza Valley, Pakistan in July 2010 (NASA Earth Observatory Image). The Hunza River flows through the section of the satellite image from north to south. At the lower end of the lake, the demolition niche, which is also shown enlarged, and the deposits of the landslide on the opposite valley slope can be seen. The overflow runs over the landslide masses through the artificial channel. The cultivated areas on the alluvial fans of the mountain streams have partially disappeared below the lake level

an initial overflow without creating a large breach to completely empty the lake through the moraine wall (Fig. 4.22).

The problem with partial erosion of the moraine wall is that ice cores and permafrost lose part of their protective sediment cover and can melt out more quickly, which leads to instability of the dam. In the regions high above the tree line, there is no stabilising effect of vegetation cover, so that once material starts to move, it might mobilise further sediments. Thus, it often happens that an initial lake emptying triggers a debris flow that continues to increase in volume down the valley and can thus have a reach as far as the city of Almaty at the foot of the mountains. To protect the villages below, a series of protective dams have been built (Fig. 4.23) to stop the debris flows (Esenov 1980; Mirtskhoulava 2000).

Similar conditions can be reported from the neighbouring valley of the Great Almatinka (Esenow 1980). On 3rd of August 1977, seepage flows in the moraine of a glacier broke open and the released water eroded a rapidly growing gully into the moraine dam. Within a few minutes, the dam crest gave way due to over-steeping of the slopes and an outburst flood developed with a peak discharge of about 210 m^3/s, while usually the discharge by seepage is about 0.8–1.0 m^3/s. The initial outburst flood mobilised additional debris in the valley bottom and along the undercut slopes, so that the water runoff developed into a debris flow. The total runoff volume increased downstream and multiplied rapidly (Fig. 4.24), while for water floods, transmission losses are characteristic (e.g. Bergman et al. 2014).

The debris flow itself reached an average thickness of 12 m, in some places with run-up effects up to 25 m. The debris flow was finally slowed down by deposition

Fig. 4.21 Usoi Dam with Lake Sarez in the Pamir Mountains, Tajikistan (NASA Earth Observatory Image)

Fig. 4.22 Terminal moraine wall of the Mametova Glacier in the Tujuksu Basin of the Tien-Shan Mountains, Kazakhstan (persons for scale). A first overflow has already eroded numerous sediments from the moraine wall and deposited them as alluvial fan below the remaining moraine dam

Fig. 4.23 Top of a series of debris flow check dams in the Little Almatinka Valley, Tien-Shan Mountains, Kazakhstan. In the haze in the background, the city of Almaty is located approximately 2000 m below

in widened valleys and by reaching a reservoir that could store its volume. Parts of the debris were remobilised by the inflowing water, resulting in about 200 local debris flows in the valley by the end of the month. As a safety measure against subsequent events, natural dammed lakes in Kazakhstan were artificially opened using heavy construction machinery so that less water can accumulate. Where this is not possible in short terms, pumps were used to reduce the volume of the lake. In case of a dam failure, the amount of water released is less and so (hopefully) the volume of debris flows is also reduced.

The individual dynamics that outbursts from moraine-dammed lakes can have are documented illustrative by

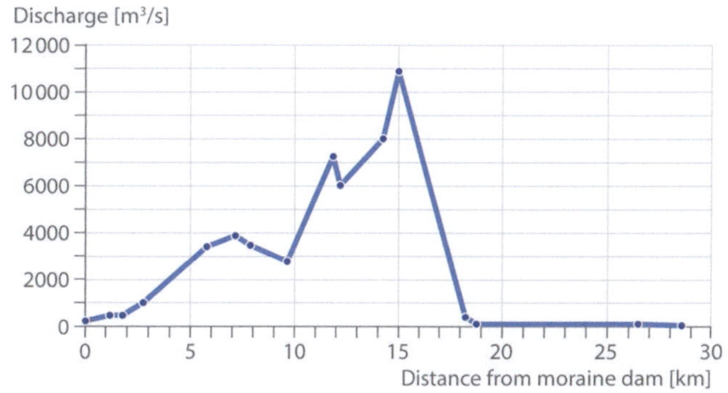

Fig. 4.24 Development of the debris flow discharge along the valley of the Great Almatinka River, Kazakhstan, on 3/4 August 1977 (data from Esenov 1980)

Kershaw et al. (2005) by a recent example from Canada. An ice avalanche of Diadem Glacier into the associated Tongue Basin Lake led to a sequence of impulse waves that overflowed the moraine embankment, eroding a channel and thus causing an outburst flood of Queen Bess Lake. Remarkably, the impulse waves overflowing the dam created a flood wave of their own in the valley below, but were subsequently caught up by the waters of the outburst flood wave released from the lake. Even though large amounts of sediment were displaced, no debris flow developed.

From the Peruvian Andes, Lliboutry et al. (1977) describe a two-stage outburst flood of a moraine-dammed lake. During the night of 16–17th of July 1951, an ice avalanche with a volume of 4–5 million m^3 fell from an overhanging glacier into Lake Artesoncocha. The resulting impulse wave flowed over the dam crest with a volume of 1.13 million m^3 and spontaneously reduced the water level in the lake by 7 m. It also eroded a spillway that merged into a waterfall on the outside of the moraine wall. Due to retrograde erosion of the waterfall edge, a second eruption finally occurred after 3 months when the waterfall had reached the lake surface. As before, the 3.52 million m^3 that spilled out flowed into another lake below, whose level was fortunately so low due to the season that there was no outburst and no danger to the town of Caraz below.

The outburst of Dig Lake in the Khumbu Himal, Nepal, was less benign (Vuichard and Zimmermann 1987). After the Langmoche Glacier had retreated from its extent during the Little Ice Age, an approximately 60 m high terminal moraine wall remained, behind which water accumulated to a water level that fluctuated according to the weather conditions. On 4th of August 1985, an ice avalanche fell from the glacier above, triggering an impulse wave estimated at around 5 m high. With the wave overtopping the dam, its erosion began. Loudly, as eye and ear witnesses reported, the outburst flood wave crashed down to the valley with an advance speed of 4–5 m/s. The outflow itself was pulsating, as at narrow points the mobilised sediments backed up themselves and the water until the water pressure was again great enough to overcome the resistance. The reconstructed hydrograph shows a steep rise within half an hour to the peak discharge of 1600 m^3/s, which then dropped again relatively quickly, so that the passage of the outburst flood wave was completed within about 5 h. The immense socio-economic damage was particularly problematic. Numerous houses were destroyed and the few agriculturally usable areas on the valley floor were eroded or covered with debris. The measures that had just begun to improve the infrastructure suffered a severe setback with the destruction of an electricity plant that was under construction. The disruption of paths, roads and bridges to the region led to a 50% increase in the price of even basic foodstuffs in the weeks following the outbreak tsunami, which illustrates the social impacts of outburst floods (cf. Carrivick and Tweed (2016) and Taylor et al. (2023) for additional aspects). But the psychological effect should not be underestimated either, because the Sherpa traditionally consider all natural events to be the work of the gods, beyond any human influence, so that one could doubt whether the construction measures had been rejected by the gods. As a consequence of the event in the summer of 1985, a systematic survey of all natural reservoirs in the region was started in order to identify potential sources of danger in advance. In the meantime, comprehensive inventories of natural reservoirs in the region are available (e.g. Mool et al. 2001a, b; Bajracharya et al. 2007).

4.3.4 Thermokarst Lakes

Thawing permafrost creates depressions and hollows that fill with groundwater, precipitation and meltwater, which in turn intensifies the melting of the frozen soil and banks (Fig. 4.25). Typically, expansion rates of 0.5 m per year are observed. What appears to be primarily scientifically interesting in the lowlands of the Arctic can become hazardous in the high mountains, because uncontrolled expansion can eventually lead to an outburst flood (Fig. 4.26), which is why a comparable thermokarst lake in the Alps has been drained by pumping (Kääb and Haeberli 2001).

Due to the small size of the lakes, it is not the resulting runoff itself that is critical, but the possibility that debris flows are introduced as a result, which can have self-reinforcing dynamics as noticed above by an example of a debris flow triggered by a moraine-dammed lake outburst flood (Fig. 4.23).

4.3.5 Lakes Dammed by Volcanic Deposits

While volcanic eruptions were previously discussed in context with the production of meltwater in ice reservoirs on Iceland, for example, and the backwater of water in caldera lakes such as Grimsvötn, now a look at the backwater effect of the deposited volcanic products themselves is taken.

Among the most spectacular lava dam failures are the events that occurred within the Grand Canyon in Arizona. During the last nearly 2 million years, volcanic eruptions have repeatedly occurred in the vicinity of what is now the Grand Canyon, damming the Colorado between 60 and 600 m high before overtopping and subsequent breach of the dam occurred (Hamblin 1994). Probably the largest outburst flood event occurred about 165,000 years ago (Fenton et al. 2006). Based on the elevation of the remaining deposits of the destroyed dam, it was possible to model the outburst flood wave and thus reconstruct the discharge conditions. An essential tool for this approach is the

Fig. 4.25 Thermokarst lake in the moraine of Soviskij Glacier, Altai Mountains, Siberia. To the right of the red arrow mark in the back, shorelines are visible, indicating a higher water level in the large lake that connected the sub-basins before drainage by an outburst flood (Fig. 4.26). The varying turbidity of the water (the uppermost lake on the left edge of the image contains clear water!) is due to a thermokarst tube connection with the turbid water containing glacial tongue basin lake above. At the front edge of the picture a fresh subsidence fissure can be seen, which proves the further subsidence of the basin. Probably due to melting of the sills between the lake basins and underground displacement of the thermokarst tubes, the basin was completely dry one year after the photo was taken, despite a wet season in-between

Fig. 4.26 Incised drainage channel of the outburst flood from the thermokarst lake above (Fig. 4.25)

geochemical identification of the dam remains dispersed by the flood wave. This is possible because the lava of different volcanic eruptions differs in terms of its chemical composition and thus the individual deposits can be differentiated. The dam at the time was 366 m high and was destroyed by overflow and subsequent incision, although its remains can still be found up to 32 km downstream. This makes this outburst flood one of the oldest flood events whose deposits can still be found today. Modelling showed that a volume of 11 km^3 of water drained within 31 h. The peak discharge was between 230,000 and 530,000 m^3/s, which exceeds the largest recent flood in the Grand Canyon by a factor of ten.

During a volcanic eruption, the volume of lava released can cause the slopes of the volcano to become unstable and descend as a landslide. Also, deposits on volcanic slopes are mobilised by meltwater that can form from snow and ice during an eruption and form thick sedimentary deposits as so-called lahars, that might block the mouth of tributaries. Waythomas (2001) describes corresponding examples from Alaska, where such lake formations have occurred repeatedly. Since the eruptions themselves could not be observed, the modelled peak discharges essentially depend on the dam-break duration that can only be assumed. In corresponding case distinctions, this results in peak discharges of several hundred thousand cubic metres for lakes with a depth of more than 150 m and volumes of several cubic kilometres.

Collapse cauldrons, which can form as calderas on the surface after the emptying and collapse of underground magma chambers, offer ideal conditions for lake basin formation. Such calderas also form at the highest altitudes of volcanic mountains high above the surrounding area. If the flanks of the volcano can no longer withstand the pressure of the ponded water, outburst floods can occur from the caldera. In his review of caldera eruptions, Manville (2011) notes that such eruption floods are among the largest flood events since the end of the last ice age. A vivid example is provided by the Aniakchak Caldera in southwestern Alaska (Waythomas et al. 1996). A 3.7 km^3 lake with an average depth of 98 m was formed in the caldera, of which shorelines, lake sediments and a small residual ponds are still preserved today (Fig. 4.27). Based on the bedding conditions, it can be concluded that between 3400 and 500 years ago, an outburst flood occurred. The site of the lake eruption is still clearly recognisable today.

Waythomas et al. (1996) applied different methods for discharge estimation. Dam breach modelling yielded peak discharges between 77,000 and 1,100,000 m^3/s depending

Fig. 4.27 Satellite image of the caldera of Aniakchak Volcano, Alaska (*Photo* NASA Earth Observer). The gap in the rim of the caldera, which points directly west, is clearly visible

on the speed of breach opening, while black-box correlation between lake or dam dimensions and discharge showed values of 80,000–280,000 m^3/s. From the size of the deposits along the flow path of the flood wave, peak discharges of 20,000–4,800,000 m^3/s can be estimated using the competence approach. The peak values refer only to the very largest blocks with diameters of up to 27 m, although there are fundamental problems in applying the competence approach in these dimensions (cf. Sect. 2.3). Finally, a peak discharge of 240,000–1,000,000 m^3/s can be estimated from the geometry of the runoff path using the Manning equation. The pros and cons of the individual approaches can be discussed at length in principle and against the background of the field evidence in detail, but it appears that a value in the six-digit range is plausible for the peak discharge.

A dramatic situation arose in the aftermath of the May 1982 eruptions of El Chichón Volcano in Mexico (Macias et al. 2004). Pyroclastic ejecta formed dams in river courses in the vicinity of the volcano, behind which the water began to impound. Barely two months after the end of the main phase of the eruption, a lake 4 km long and 300–400 m wide with a volume of 40,000,000 m^3 was formed in the valley of Magdalena River. Due to the contact with the still hot volcanic deposits, the lake water still had temperatures of around 90° Celsius even during the following outburst flood. When the dam failed in the night of 26th of May 1982 due to overflowing, about 11,000 m^3 of almost boiling water poured into the valley below. This outburst flood could be stopped in an artificial reservoir for hydroelectric power generation, which abruptly rose by 7 m in the process. Unfortunately, the still 50–52 °C warm water claimed more victims weeks after the volcanic eruption, which had already killed almost 2000 people.

Lake outburst floods following backwater caused by volcanic deposits can also be observed in Central Europe. When the Laacher See Volcano, located in the south-eastern Eifel Mountains in western Germany, erupted about 12,900 years ago, pyroclastic flows rushed over the slopes and through the valleys towards the northern Middle Rhine Valley (Baales et al. 2002; Park and Schmincke 1997). The deposited pumice masses are lighter than water due to the air pockets they contain, so that they floated on the River Rhine. Due to the masses of pumice, however, the river was soon no longer able to discharge all the sediment and began to back up. Deposits of pumice that had been relocated by flowing water (Fig. 4.28) were found up to an elevation of 85 m, which corresponds to a lake depth of up to 27 m in some places along the river valley.

In detail, evidence for two dams is determined, one of them at the mouth of the Brohl Creek valley, which acted

Fig. 4.28 Pumice of the Laacher See Volcano in an outcrop in the Neuwied Basin, River Rhein at Koblenz (*Photo* Giles Erkens)

as a pathway for the pyroclastic flows. The other formed in the area of the village of Andernach upstream the narrowing of the Rhine Valley after flowing through the Neuwied Basin. The backwater of the River Rhine and its tributaries at a height of about 85 m led to the formation of a reservoir of about 2.5 km^3, which probably reached as far upstream as the city of Mainz. Since the corresponding deposits were deposited within the recent floodplain area of the Rhine, no obvious field evidence has been preserved. Since the dams were formed by floating deposits, they will not have been completely impermeable, so that an impoundment may have taken longer times. For the deposits at the mouth of the Brohl Creek valley, however, Park and Schmincke (2009) assume a massive dam, which is supported by the dimensions of the deposits and their location in the straight pathway of the pyroclastic flow. Eventually, the deposits could withstand the pressure of the water for a limited time and an outburst flood poured through the northern Middle Rhine Valley. Traces of corresponding pumice-bearing deposits have been found as far downstream as Bonn. Since the volcanism of the Eifel Mountains is currently only dormant, not extinct, volcanic eruptions in the region cannot be ruled out for the future. What would happen if a reservoir were to form today as a result of a volcanic eruption has been imagined by Schreiber (2008) in a novel that is well worth reading due to the scientific background of the author as professor of geology.

4.3.6 Artificial Dammed Lakes

Unfortunately, not all attempts to impound water by artificial dams have been permanently successful, resulting in dam failures and outburst floods as well. Overviews with a variety of engineering dam parameters are provided e.g. by Singh (1996), Wahl (1998) and Wu (2011) while at https://damfailures.org/ the Association of State Dam Safety Officials provides an online-database on "Lessons learned from dam failures and incidents".

A spectacular dam failure occurred on 2nd of December 1959 at the Malpasset Dam above the small town of Frejus in southern France (Londe 1987; Singh 1996). The double arch dam, completed in 1954 to manage water for drinking and irrigation purposes, had not yet been dammed to its full height when heavy rainfall at the end of November caused the water level in the lake to rise rapidly. Considerations to reduce the impoundment by opening the outlets were initially delayed so as not to endanger the motorway construction site below. In the evening, the entire dam abruptly gave way without further warning in the form of a twisting movement, so that an initially 40 m high wave poured into the valley of the Reyran Creek. Two villages below were completely submerged by the water, which cost a total of 421 people their lives. The wave was still 3 m high in Frejus, 12 km away, which was reached after 21 min. Subsequent investigations revealed that on one side of the valley the abutment of the arch dam could not withstand the pressure and thus the entire contents of the lake suddenly drained.

On 9th of October 1963, a landslide occurred on the flank of Monte Toc in the Italian Vajont Valley in the Carnic Alps, when around 250 million m^3 of rock fell into the reservoir below (Müller 1964; Genevois and Ghirotti 2005; Paronuzzi and Bolla 2015). The lake, which was only 2/3 full with 115 million m^3 of water, was largely filled in and the water it contained was displaced by the landslide. The impulse wave rose 260 m high on the opposite slope and then plunged 400 m down to the valley. Five villages below were destroyed, killing more than 1900 people. Along the path of the impulse wave, the rock was eroded several metres deep in places. The 267 m high dam itself remained intact and still stands in the valley today while the entire storage basin is filled by the landslide (Fig. 4.29).

The failure of the Teton Dam in the US state of Idaho on 5th of June 1976 is well documented (Chadwick et al. 1976; Ray et al. 1978; Seed and Duncan 1987). Water seeping through the earthen dam escaped on the air side about 30 m below the dam crest. The outlet rapidly increased in size due to erosion and incision downstream, moving backwards towards the top of the dam, which then collapsed (Fig. 4.30).

Half an hour after the first water discharge, the flow had already increased to such an extent that a gradually growing rotating vortex of water flowing through the dam formed on the lake surface. After two hours, the retrograde erosion had reached the top of the dam, which broke open shortly afterwards. Only 5 h after the first water release, practically the entire reservoir with a water content of 308 million m^3 had spilled out. The outburst flood reached a peak discharge of 65,000 m^3/s and cost 14 people their lives. The cause of the development of the seepage flow was found to be insufficient preliminary geological exploration of the open fissures in the adjacent rock and inadequate compaction of the dam materials.

In connection with historical flood events in China, reference has already been made to dam openings as a tactical means of warfare (cf. Sect. 3.1). This was also used during the Second World War, when the British Royal Air Force attacked dams in Germany with specially developed rolling bombs (Herget and Gregori 2020). The targets of the attacks were reservoirs in the Sauerland region with the background of both causing destruction by the outburst flood waves and disrupting the drinking water supply in the Ruhr District area, the key armoury of Nazi-Germany. After several unsuccessful attempts, the Möhne and Eder Dams were finally hit and destroyed in May 1943 (Fig. 4.31).

Fig. 4.29 Intact dam in the Vajont River valley with the landslide mass filling the lake visible in the background

Fig. 4.30 Gap in Teton Dam, Idaho, in June 1976

After the successful bombardment, a small crack initially appeared at the Möhne Dam, but the water flowing through it quickly enlarged it to a width of 77 m and a depth of 22 m. Initially, an estimated 9000 m^3/s, reducing quickly to 6000 m^3/s drained. The flood wave reached the village of Neheim, located in the River Ruhr valley, after almost 30 min with a height of 12 m and a flow velocity of about 6 m/s. In total, more than 1200 people died. A total of over 1200 people died, most of them in prison camps located in the floodplain of Möhne Creek at Neheim. As far as the central Ruhr

Fig. 4.31 Destroyed Möhne Dam and mostly drained reservoir on the day after the bombing in May 1943

District, the outburst flood wave led to unprecedented water levels, flooding waterworks and clogging filter basins with mud, as well as destroying numerous bridges. The situation was similar in the area below the Eder Dam, which was attacked in the same raid. Fortunately, the damage was less severe here because of the less dense infrastructure along the valley bottom of the River Eder. development. Anyhow, in Hannoversch Münden, located about 70 km downstream of Eder Dam the flood level was among the largest of the historic record (Fig. 3.19).

During the ongoing offensive war of Russia against Ukraine man-made dams were twice damaged. In the beginning of the war, large parts of the country northwest of Kiev where inundated successfully to defend the capital against the attacking Russian troops. On 6th of June 2023, Kakhovka Dam located in the Dnipro River valley upstream of the city of Kherson was destroyed. It was part of a cascade of hydroelectric dams along the river and had a height of 30 m and length of 3840 m (UN EP 2023). Due to missing accessibility only remote sensing and modelling approaches could be supplied for the flood estimation: 30,000–50,000 m^3/s for the dam location, 25,000–35,000 m^3/s at Kherson city and 23,000–32,000 m^3/s at the Dnipro estuary. Considering the mean discharge of 1680 m^3/s at the dam location, the flood lasting 2 weeks probably had a magnitude of a statistical return period of > 10,000 years.

References

Alford D, Schuster R (eds) (2000) Usoi landslide dam and Lake Sarez—an assessment of hazard and risk in the Pamir Mountains, Tajikistan. United Nations International Strategy for Disaster Reduction Prevention Series 1.

Alho P, Russell AJ, Carrivick JL, Käyhkö J (2005) Reconstruction of the largest Holocene jökulhlaup within Jökulsa a Fjöllum, NE Iceland. Quatern Sci Rev 24:2319–2334

Baales M, Jöris O, Street M, Bittmann F, Weninger B, Wietholt J (2002) Impact of the late glacial eruption of the Laacher See Volcano, Central Rhineland, Germany. Quatern Res 58:273–288

Bajracharya SR, Mool PK, Shrestha BR (2007) Impact of climate change on Himalayan glaciers and glacial lakes. Kathmandu (ICIDMOD).

Bergman N, Sholker O, Roskin J, Greenbaum N (2014) The Nahal Oz Reservoir dam-break flood - geomorphic impact on a small ephemeral loess-channel in the semi-arid Negev Desert, Israel. Geomorphology 210:83–97

Björnsson H (2002) Subglacial lakes and Jökulhlaups in Iceland. Global Planet Change 35:255–271

Björnsson H (2009) Jökulhlaups in Iceland—sources, release and drainage. In: Burr DM, Carling PA, Baker VR (eds) Megaflooding on Earth and Mars. Cambridge (Cambridge University Press), pp 50–64

Björnsson H (2010) Understanding jökulhlaups - from tale to theory. J Glaciol 56:1002–1010

Björnsson H (2016) The glaciers of Iceland—a historical, cultural and scientific overview. Berlin (Springer)

Bussmann F, Anselmetti FS (2010) Rossberg landslide history and flood chronology as recorded in Lake Lauerz sediments (Central Switzerland). Swiss Journal of Geoscience 103:43–59

Butler DR, Malanson GP (2005) The geomorphic influences of beaver dams and failures of beaver dams. Geomorphology 71:48–60

Carrivick JL, Russell AJ, Tweed FS, Twigg D (2004) Palaeohydrology and sedimentary impacts of jökulhlaups from Kverkfjöll, Iceland. Sed Geol 172:19–40

Carrivick JL, Jones R, Keevil G (2011) Experimental insights on geomorphological processes within dam break outburst floods. J Hydrol 408:153–163

Chadwick WL, Casagrande A, Dowd HA, Fucik EM, Higginson RK, Leps TM, Peck RB, Seed HB, Jansen RB (1976) Report to US department of the interior and state of idaho on failure of Teton Dam. Washington (US Government Printing Office)

Chanson H (2005) The 1786 earthquake-triggered landslide dam and subsequent dam-break flood on the Dadu-River, southwestern China. Geomorphology 71:437–440

Clague JJ, Evans SG (2000) A review of catastrophic drainage of moraine-dammed lakes in British Columbia. Quatern Sci Rev 19:1763–1783

Clarke GKC (2003) Hydraulics of subglacial outburst floods - new insights from the Spring-Hutter formulation. J Glaciol 49:299–313

Cook SJ, Quincey DJ (2015) Estimating the volume of Alpine glacial lakes. Earth Surf Dyn 3:559–575

Costa JE, Schuster RL (1988) The formation and failure of natural dams. Geol Soc Am Bull 100:1054–1068

Costa JE (1988) Floods from dam failures. In: Baker VR, Kochel RC, Patton PC (eds) Flood geomorphology. New York (Wiley), pp 439–463
Dai FC, Lee CF, Deng JH, Tham LG (2005) The 1786 earthquake-triggered landslide dam and subsequent dam-break flood on the Dadu-River, southwestern China. Geomorphology 65:205–221
De Groot MB, den Adel H, Stourjesdijk TP, van Westenbrugge CJ (1995) Risk of dike failure due to flow slides. Coastal Eng 26:241–249
Delaney K, Evans S (2017) The evolution (2010–2015) and engineering mitigation of a rockslide-dammed lake (Hunza River, Pakistan); characterisation by analytical remote sensing. Eng Geol 220:65–75
Deplazes G, Anselmetti FS, Hajdas I (2007) Lake sediments deposited on the Flims rockslide mass - the key to date the largest mass movement of the Alps. Terra Nova 19:252–258
Erlingsson U (2006) Lake Vostok behaves like a "captured lake" and may be near to creating an Antarctic jökulhlaup. Geographiska Annaler 88A:1–7
Esenow UE (1980) Der Murgang vom Jahre 1977 im Wildfluß "Große Almatinka" und das Schutzproblem der Stadt Alma-Ata. Interpraevent Publikationen 4:245–249
Fenton CR, Webb RH, Cerling TE (2006) Peak discharge of a Pleistocene lava-dam outburst flood in Grand Canyon, Arizona, USA. Quatern Res 65:324–335
Fritz HM, Hager WH, Minor H-E (2001) Lituya Bay case - rockslide impact and wave run-up. Sci Tsunami Haz 19:3–22
Genevois R, Ghirotti M (2005) The 1963 Vaiont landslide. Giornale Di Geologia Applicata 1:41–52
Gore DB (1992) Ice-damming and fluvial erosion in the Vestfold Hills, East Antarctica. Antarct Sci 4:227–234
Haeberli W (1983) Frequency and characteristics of glacier floods in the Swiss Alps. Ann Glaciol 4:85–90
Hamblin WK (1994) Late Cenozoic lava dams in the western Grand Canyon. Geological Society of America Memoir 183
Heine K (2004) Flood reconstructions in the Namib Desert, Namibia and Little Ice Age climatic implications—evidence from slack-water deposits and desert soil sequences. J Geol Soc India 64:535–547
Herget J, Gregori L (2020) Outburst flood from Möhne Reservoir in May 1943 after aerial bombing. In: Herget J, Fontana A (eds) Palaeohydrology—traces, tracks and trails of extreme events. Cham (Springer), pp 49–61
Herget J, Schütte F, Klosterhalfen A (2015) Empirical modelling of outburst flood hydrographs. Z Geomorphol 59 (Suppl. 3):177–198
Hermanns RL, Niedermann S, Ivy-Ochs S, Kubik PW (2004) Rock avalanching into a landslide-dammed lake causing multiple dam failures in Las Conchas valley (NW Argentina) - evidence from surface exposure dating and stratigraphic analyses. Landslides 1:113–122
Hewitt K (1998) Catastrophic landslides and their effects on the Upper Indus streams, Karakoram Himalaya, northern Pakistan. Geomorphology 26:47–80
Holzhauser H (1997) Fluctuations of the Grosser Aletsch Glacier and the Gorner Glacier during the last 3200 years—new results. In: Frenzel B (ed) Glacier fluctuations during the Holocene. Stuttgart (Fischer) pp 35–58
Hormes A, Müller BU, Schlüchter C (2001) The Alps with little ice—evidence for eight Holocene phases of reduced glacier extent in the Central Swiss Alps. The Holocene 11:255–265
Hormes A, Schlüchter C, Stocker TF (1998) Minimal extension phases of Unteraarglacier (Swiss Alps) during the Holocene based on 14C analysis of wood. Radiocarbon 40:809–817
Huggel C, Kääb A, Haeberli W, Teysseire P, Paul F (2002) Remote sensing based assessment of hazards from glacier lake outburst—A case study in the Swiss Alps. Can Geotech J 39:316–330
Iken A, Röthlisberger H, Flotron A, Haeberli W (1983) The uplift of Unteraargletscher at the beginning of the melt season - a consequence of water storage at the bed? J Glaciol 29:28–47
Ivy-Ochs S, Poschinger AV, Synal H-A, Maisch M (2009) Surface exposure dating of the Flims landslide, Graubünden, Switzerland. Geomorphology 103:104–112
Ivy-Ochs S, Kerschner H, Maisch M, Christl M, Kubik PW, Schlüchter C (2009) Latest Pleistocene and Holocene glacier variations in the European Alps. Quatern Sci Rev 28:2137–2149
Kääb A, Haeberli W (2001) Evolution of a high-mountain thermokarst lake in the Swiss Alps. Arct Antarct Alp Res 33:385–390
Kershaw JA, Clague JJ, Evans SG (2005) Geomorphic and sedimentological signature of a two-phase outburst flood from moraine-dammed Quee Bess Lake, British Columbia, Canada. Earth Surf Proc Land 30:1–25
King L (1987^2) Studien zur postglazialen Gletscher- und Vegetationsgeschichte des Sustenpassgebietes. Giessen (Geographisches Institut der Universität)
Korup O, Tweed F (2007) Ice, moraine, and landslide dams in mountainous terrain. Quatern Sci Rev 26:3406–3422
Krainer K, Spieler A (1999) The sedimentary record of ice-dammed lakes in the Ötztal Alps Austria. Zeitschrift Für Gletscherkunde u. Glazialgeologie 35:65–86
Lliboutry L, Morales Arnao B, Pautre A, Schneider B (1977) Glaciological problems set by the control of dangerous lakes in Cordillera Blanca, Peru - I. Historical failures of morainic dams, their causes and prevention. J Glaciol 18:239–254
Londe P (1987) The Malpasset dam failure. Eng Geol 24:295–329
Lütschg O (1915) Der Märjelensee und seine Abflussverhältnisse - eine hydrologische Studie unter Mitberücksichtigung hydrographischer Erscheinungen in anderen Flussgebieten. Annalen der Schweizer Landeshydrographie 1.
Macias JL, Capra L, Scott KM, Espinola JM, Garcia-Palomo A, Costa JE (2004) The 26 May 1982 breakout flows derived from failure of a volcanic dam at El Chichón, Chiapas, Mexico. Geol Soc Am Bull 116:233–246
Manville V (2011) An overview of break-out floods from intracaldera lakes. Global Planet Change 70:14–23
Mason K (1929) Indus floods and Shyok Glaciers. Himalayan Journal 1:10–29
Mayo LR (1989) Advance of Hubbard Glacier and 1986 outburst of Russell Fiord, Alaska, USA. Ann Glaciol 13:189–194
Miller DJ (1960) Giant waves in Lituya Bay, Alaska. US Geological Survey Professional Paper 354-C.
Mirtskhoulava TE (2000) Mud floods and their prediction in the Caucasus and Central Asia. In: Parker DJ (ed) Floods—Volume II. London (Routledge), 216–222.
Molnia BF (2008) Glaciers of Alaska. US Geological Survey Professional Paper 1386-K
Mool PK, Wangda D, Bajracharya SR, Kunzang K, Gurung DR, Joshi SP (2001a): Inventory of glaciers, glacial lakes and glacial lake outburst floods, Bhutan. Kathmandu (ICIMOD)
Mool PK, Bajracharya SR, Joshi SP (2001b) Inventory of glaciers, glacial lakes and glacial lake outburst floods, Nepal. Kathmandu (ICIMOD)
Morche D, Schmidt K-H, Heckmann T, Haas F (2007) Hydrology and geomorphic effects of a high-magnitude flood in an alpine river. Geogr Ann 89A:5–19
Morris M, Hanson G, Hassan M (2008) Improving the accuracy of breach modelling—why are we not progressing faster? Journal of Flood Risk Management 1:150–161
Motyka R, Lawson D, Finnegan D, Kalli G, Molnia B, Arendt A (2008) Hubbard Glacier update—another closure of Russell Fiord in the making? J Glaciol 54:562–564
Müller L (1964) The rock slide in the Vajont valley. Felsmechanik und Ingenieurgeologie 2:148–212

NGO = Naturforschende Gesellschaft Oberwallis (ed) (1995) Der Bergsturz von Randa 1991. Brig (Tscherrig).
Nicolussi K (2013) Die historischen Vorstöße und Hochstände des Vernagtferners 1600–1850 AD. Z Gletscherk Glazialgeol 45(46):9–23
O'Connor JE, Beebee RA (2009) Floods from natural rock-material dams. In: Burr DM, Carling PA, Baker VR (eds) Megaflooding on Earth and Mars. Cambridge (Cambridge University Press), pp 128–171
Park C, Schmincke H-U (1997) Lake formation and catastrophic dam burst during the Late Pleistocene Laacher See eruption (Germany). Naturwissenschaften 84:521–525
Park C, Schmincke H-U (2009) Apokalypse im Rheintal. Spektrum der Wissenschaft 2(2009):78–87
Paronuzzi P, Bolla A (2015) Gravity-induced rock mass damage related to large en masse rockslides—evidence from Vajont. Geomorphology 234:28–53
Patzelt G (1994) Former lakes dammed by glaciers and resulting floods in the Ötztal, Tyrol. Mt Res Dev 14:298–301
Petley D (2010) Hunza landslide monitoring—webside displaying graphs and data on the current state of the Attabad landslide dam in Hunza, N. Pakistan. http://hunzalandslide.blogspot.com/ (28 May 2024)
Petley D (2011) Damming events at Attabad. International Water Power and Dam Construction 21 March 2011. https://www.waterpowermagazine.com/features/featuredamming-events-at-attabad-/
Post AS (1967) Effects of the March 1964 Alaskan earthquake on glaciers. US Geological Survey Professional Papers 544-D
Ray HA, Kjelstrom LC, Crosthwaite EG, Low WH (1978) The flood in southeastern Idaho from the Teton Dam failure of June 5, 1976. US Geological Survey Open File Report 77–765
Raymond CF, Nolan M (2000) Drainage of a glacial lake trough an ice spillway. In: Nakawo MCF, Raymond, Fountain AG (eds) Debris-covered glaciers. IAHS-Publication 264, Wallingford (IAHS), pp 199–207
Richardson SD, Reynolds JM (2000) An overview of glacial hazards in the Himalayas. Quatern Int 65(66):31–47
Risley J, Walder J, Denlinger R (2006) Usoi dam wave overtopping and flood routing in the Bartang and Panj Rivers, Tajikistan. US Geological Survey Water-Resources Investigations Report 03–4004.
Roberts MJ (2005) Jökulhlaups—A reassessment of floodwater flow through glaciers. Rev Geophys 43:RG1002. https://doi.org/10.1029/2003RG000147
Russell AJ, Gregory AR, Large ARG, Fleisher PJ, Harris TD (2007) Tunnel channel formation during the November 1996 jökulhlaup, Skeidararjökull, Iceland. Ann Glaciol 45:95–103
Russell AJ, Roberts MJ, Fay H, Marren PM, Cassidy NJ, Tweed FS, Harris T (2006) Icelandic jökulhlaup impacts - implications for ice-sheet hydrology, sediment transfer and geomorphology. Geomorphology 75:33–64
Schindler C, Cuenod Y, Eisenlohr T, Joris C-L (1993) Die Ereignisse vom 18. April und 9. Mai 1991 bei Randa (VS) - ein atypischer Bergsturz in Raten. Eclogae Geologica Helvetica 86:643–665
Schlüchter C, Jörin U (2004) Alpen ohne Gletscher? Die Alpen 6(2004):34–47
Schneider J-L, Pollet N, Chapron E, Wessels M, Wassmer P (2004) Signature of Rhine Valley sturzstrom dam failures in Holocene sediments of Lake Constance, Germany. Sed Geol 169:75–91
Schreiber UC (2008) Die Flucht der Ameisen—ein Geo-Thriller. München (Piper)
Seed HB, Duncan JM (1987) The failure of Teton Dam. Eng Geol 24:173–205
Siegert MJ (2005) Reviewing the origin of subglacial Lake Vostok and its sensitivity to ice sheet changes. Prog Phys Geogr 29:156–170
Siegert MJ, Carter S, Tabacco I, Popov S, Blankenship DD (2005) A revised inventory of Antarctic subglacial lakes. Antarct Sci 17:453–460
Smith N (1971) A history of dams. Secaucus (Citadel Press)
Stearns LA, Hamilton GS, van der Veen CJ, Finnegan DC, O'Neel S, Scheick JB, Lawson DE (2015) Glaciological and marine geological controls on terminus dynamics of Hubbard Glacier, southeast Alaska. J Geophys Res 120:1065–1081
Stone R (2009) Peril in the Pamirs. Science 326:1614–1617
Svendsen J, Stollhofen H, Krapf CBE, Stanistreet IG (2003) Mass and hyperconcentrated flow deposits record dune damming and catastrophic breakthrough of ephemeral rivers, Skeleton Coast Erg, Namibia. Sed Geol 160:7–31
Tomasson H (1996) The jökulhlaup from Katla in 1918. Ann Glaciol 22:249–254
Trabant DC, March RS, Thomas DS (2003) Hubbard Glacier, Alaska: growing and advancing in spite of global climate change and the 1986 and 2002 Russell Lake outburst floods. US Geological Survey Fact Sheet 001–03.
Tweed FS, Russell AJ (1999) Controls on the formation and sudden drainage of glacier-impounded lakes - implications for jökulhlaup characteristics. Prog Phys Geogr 23:79–110
Tweed FS (2000) Jökulhlaup initiation by ice-dam flotation - the significance of glacier debris content. Earth Surf Proc Land 25:105–108
UN EP = United Nations Environment Programme (2023) Rapid Environmental Assessment of Kakhovka Dam Breach, Ukraine 2023. Nairobi (UN EP). 10.59117/20.500.11822/43696
UN-ISDR – UN International Strategy of Disaster Reduction Inter-Agency Secretariat (ed) (2007): Sarez Lake—the latest achievements and unsolved problems. Dushanbe
Viollet P-L (2007) Water engineering in ancient civilizations - 5,000 years of history. Madrid (International Association of Hydraulic Engineering and Research)
von Poschinger A (2005) Der Flimser Bergsturz als Staudamm. Bulletin Für Angewandte Geologie 10:33–47
Vuichard D, Zimmermann M (1987) The 1985 catastrophic drainage of a moraine-dammed lake, Khumbu Himal, Nepal—cause and consequence. Mt Res Dev 7:91–110
Wahl TL (1998) Prediction of embankment dam breach parameters - a literature review and needs assessment. Dam Safety Office des Water Resources Laboratory im US Department of Interior, DSO-98-004.
Wahl TL (2004) Uncertainty of predictions of embankment dam breach parameters. J Hydraul Eng 130:389–397
Walder JS, Costa JE (1996) Outburst floods from glacier-dammed lakes - the effect of mode of lake drainage on flood magnitude. Earth Surf Proc Land 21:701–723
Wassmer P, Schneider JL, Pollet N, Schmitter-Voirin C (2004) Effects of the internal structure of a rock-avalanche dam on the drainage mechanism of its impoundment, Flims sturzstrom and Ilanz palaeo-lake, Swiss Alps. Geomorphology 61:3–17
Waythomas CF (2001) Formation and failure of volcanic debris dams in the Chakachatna River valley associated with eruptions of the Spurr volcanic complex, Alaska. Geomorphology 39:111–129
Waythomas CF, Walder JS, McGimsey RG, Neal CA (1996) A catastrophic flood caused by drainage of a caldera lake at Aniakchak Volcano, Alaska, and implications for volcanic hazards assessment. Geol Soc Am Bull 108:861–871
Wingham DJ, Siegert MJ, Shepherd A, Muir AS (2006) Rapid discharge connects Antarctic subglacial lakes. Nature 440:1033–1036
Worni R, Huggel C, Clague JJ, Schaub Y, Stoffel M (2014) Coupling glacial lake impact, dam breach, and flood processes - a modeling perspective. Geomorphology 224:161–176

Further Reading

The listed review papers on the formation and failure of natural dams provide numerous additional hints and perspectives:

Carrivick JL Tweed FS (2019) A review of glacier outburst floods in Iceland and Greenland with a megafloods perspective. Earth-Science Reviews 196:102876

Cenderelli DA (2000) Floods from natural and artificial dam failures. In: Wohl EE (ed) Inland flood hazards. Cambridge (Cambridge University Press), pp 73–103

Costa JE, Schuster RL (1988) The formation and failure of natural dams. Geol Soc Am Bull 100:1054–1068

Costa JE (1988) Floods from dam failures. In: Baker VR, Kochel RC, PC Patton (eds) Flood geomorphology. New York (Wiley), pp 439–463

Korup O, Tweed F (2007) Ice, moraine, and landslide dams in mountainous terrain. Quatern Sci Rev 26:3406–3422

Liu W, Carling PA, Hu K, Wang H, Zhou Z, Zhou L, Liu D, Laig Z, Zhang X (2019) Outburst floods in China—a review. Earth Sci Rev 197:102895

O'Connor JE, Beebee RA (2009) Floods from natural rock-material dams. In: Burr DM, Carling PA, Baker VR (eds) Megaflooding on Earth and Mars. Cambridge (Cambridge University Press), pp 128–171

O'Connor JE, Grant GE, Costa JE (2002) The geology and geography of floods. In: House PK, Webb RH, Baker VR, Levish DR (eds) Ancient floods, modern hazards—principles and applications of palaeoflood hydrology. Washington (American Geophysical Union), pp 359–385

Borodavko PS, Glazirin GE, Herget J, Severskiy IV (eds) (2013) Hazard assessment and outburst flood estimation of naturally dammed lakes in Central Asia. Aachen (Shaker)

Fountain AG, Walder JS (1998) Water flow through temperate glaciers. Rev Geophys 36:299–328

Glazirin GE (2013) Key locations of naturally dammed lakes in Uzbekistan. In: Borodavko PS, Glazirin GE, Herget J, Severskiy IV(eds) Hazard assessment and outburst flood estimation of naturally dammed lakes in Central Asia. Aachen (Shaker), pp 3–20

ICOLD - International Commission On Large Dams (ed) (1998) Dam-break flood analysis. ICOLD Bulletin 111

Manville V (2001) Techniques for evaluating the size of potential dam-break floods from natural dams. Institute of Geological and Nuclear Science - science report 2001/28.

Schuster RL (ed) (1986) Landslide dams—processes, risk, and mitigation. American Society Civil Engineering Geotechnical Special Publication 3.

Singh VP (1996) Dam breach modelling technology. Dordrecht (Kluwer).

Vischer DL, Hager WH (1998) Dam hydraulics. Chichester (Wiley)

Westoby MJ, Glasser NF, Brasington J, Hambrey MJ, Quincey DJ, Reynolds JM (2014) Modelling outburst floods from moraine-dammed glacial lakes. Earth-Sci Rev 134:137–159

Wu W (ed) (2011) Earthen embankment breaching. J Hydraulic Eng 137:1549–1564

A state of the art paper including research perspectives is presented by

Emmer A 36 co-authors (2022) Progress and challenges in glacial lake outburst flood research (2017–2021)—a research community perspective. Natural Hazards and Earth System Sciences 22:3041–3061

A statistical analysis about frequencies and impacts of glacial lake outburst floods is presented by

Carrivick JL, Tweed FS (2016) A global assessment of the societal impacts of glacier outburst floods. Global Planet Change 144:1–16

In addition to previous review papers mentioned above among the cited references, several databases are established. Some are with a designated regional focus, some are of global focus respectively available online:

Veh G, Lützow N (2024) Glacier lake outburst flood database V4.0. http://glofs.geoecology.uni-potsdam.de/ (28 May 2024)

Borodavko PS, Glazirin GE, Herget J, Severskiy IV (eds) (2013) Hazard assessment and outburst flood estimation of naturally dammed lakes in Central Asia. Aachen (Shaker)

International Centre for Integrated Mountain Development (ed.) (2023): GLOF database of High Mountain Asia. 10.26066/RDS.1973283

Norwegian Water Resources and Energy Directorate (ed.): Glacier lake outburst floods. https://glacier.nve.no/Glacier/viewer/GLOF/en/ (12 Feb 2024)

Emmer, A. and 26 co-authors (2022) 160 glacial lake outburst floods (GLOFs) across the Tropical Andes since the Little Ice Age. Global and Planetary Change 208. 10.1016/j.gloplacha.2021.103722

Hazard estimations by glacial lake outburst floods are carried out on global scale e.g. by:

Taylor C, Robinson TR, Dunning S, Carr JR, Westoby M (2023) Glacial lake outburst floods threaten millions globally. Nat Commun 14:487

Wahl TL (1998) Prediction of embankment dam breach parameters - a literature review and needs assessment. Dam Safety Office of the Water Resources Laboratory of the US Department of Interior, DSO-98-004

Megafloods in Pleistocene Times

5

The recent history of the Earth over the last 2.7 million years is characterised by a repeated periodic alternation of ice ages and warm periods (e.g. Cohen and Gibbard 2019 and 2022, Ehlers 2022). Within the framework of the constantly changing plate tectonic constellation of the distribution of continents, mountains, land bridges and ocean basins, the Earth is able to react to the periodic fluctuations of the earth's orbital elements around the sun through global cooling and re-warming. As a result, snow remains regionally throughout the year, develops further into ice, which eventually begins to flow and expand. This is how the large ice sheets that have shaped Central Europe from the north from Scandinavia, for example, and in the form of the valley glacier systems emerging from the Alps, were generated. Large areas of the earth, especially all high-mountain ranges were thus influenced by advanced ice masses during the Pleistocene (Ehlers and Gibbard 2004; Ehlers et al. 2011), as the period of repeated alternations of ice and warm periods is called.

In particular, it were the large continental ice sheets that had a lasting effect on the river network. They blocked river courses and thus dammed lakes filling up entire valley systems, diverted water bodies or directed the surface flow along more or less temporary ice-marginal valley systems (Baker and Carling 2020). Some of the lakes were continental in extent and their outburst floods completely reshaped the entire landscape. This fundamental change of the relief is typical of the lasting consequences of Pleistocene outburst floods and is characterized as cataclysmic (Abele 1981; Baker 1988). The enormous water masses, which have exceeded recent and historical floods by several orders of magnitude, have in part also had global climatic consequences, even though their influence on the post-glacial sea level rise was of minor significance (Harrison et al. 2018).

5.1 Ice-Marginal Drainage Dynamics in Europe

Central Europe was largely shaped during the ice ages by the Nordic Ice Sheet originating from Scandinavia and the glaciation of the mountains, namely the Alps. The glaciation of the mountains—in Germany, besides the Alps, the Harz, the Black Forest and the Bavarian Forest were affected—had relatively little influence on the river networks (cf. Ehlers and Gibbard 2004; Ehlers et al. 2011 for other mountain ranges in Europe). The glaciers followed the valley lines out of the Alps into the foothills and their meltwater drained through the existing valleys. There were individual ice-dammed lakes in the mountains, but these were only of local significance. The situation was different at the northern margin of the low mountain ranges (Liedtke 1981). The Nordic Ice Sheet that advanced up to here moved against the gradient of the rivers draining northwards from the low mountain ranges to the Baltic and North Seas and dammed them up. From the third last ice age, the Elsterian Ice Age, which lasted from 400,000 to 320,000 years before today, numerous ice-dammed lakes have survived in eastern Germany with their preserved characteristic varved clay sediments (Junge 1998). The remains of the last two ice ages, the Saalian Ice Age (300,000–128,000 years before present) and the Weichselian Ice Age (115,000–10,000 years before present), are much richer.

During the Saalian Ice Age, the Nordic Ice Sheet advanced as far as the Lower Rhine at the region of the city of Düsseldorf. The River Rhine was also crossed in the process and shifted westwards by the ice lobe (Busschers et al. 2008; Skupin and Zandstra 2011). Something similar occurred at the southern margin of the Münsterland Embayment, where the ice shield dammed up the River Ems and also blocked the path of the River Lippe, which was draining to the west. Temporary drainage courses

J. Herget, *Flood Reconstruction*, Springer Natural Hazards, https://doi.org/10.1007/978-3-031-66815-9_5

developed parallel to the advancing ice margin, the deposits of which have been preserved at different elevations (Herget 2002). In numerous places, ice-dammed lakes were formed in front of the ice margin, although the validity of the findings on some of the lakes is controversial. In considerations of the dynamics of the lakes, overflows into adjacent catchments played an important role (Thome 1983; Herget 1998), while reliable evidence of outburst floods was not found. Meinsen et al. (2011) analysed evidence of outburst floods from the Saalian-age ice-dammed lake in the River Weser valley. The lake was formed when the Nordic Ice Sheet reached the Weserbergland- and Wiehen-Mountains, where it blocked the outlet of the River Weser from the southern adjacent low mountain ranges. Based on the location and structure of the preserved deposits, the development of the lakes can be traced in detail, taking into account the topography (Winsemann et al. 2011, 2015; Winsemann and Lang 2020). During the maximum ice extent, in addition to the lake in the River Weser valley, another lake existed on the Paderborn Plateau in the south-east of the Münsterland Embayment. This lake had a spillway into the adjacent Möhne and Ruhr Valleys to the south at an altitude of about 350 m, which were not reached by the advancing ice masses because they were held back here by the ridge of the Haarstrang escarpment (Fig. 5.1).

During the advance of the ice-shield, various spillways existed from the lake in the Weser Valley to the one in the Münsterland Embayment. When the maximum ice extent was reached, these were also blocked and Lake Weser dammed up to a height of 200 m above sea level. As part of a climatic fluctuation during the Saalian Ice Age, the lobe of the inland ice began to melt down and the active ice margin was displaced. In this process, the numerous passes through the Teutoburger Wald mountain range (Fig. 5.1) became active as spillways and water poured into the Münsterland Embayment. According to the elevation of the spillways, the water flowed through the Teutoburger Wald at different points. In a first lowering to a level of 135 m, about 90 km^3 of water were thus released, in a second phase another 20 km^3 of water. Streamlined hills, changes in sediment composition and scouring below the spillways on the southern side of the Teutoburger Wald mountain range are considered as indicators of the outburst floods and interpreted hydraulically. The flows through the individual spillways, which can be differentiated temporally and spatially, had peak values between 250,000–1,300,000 m^3/s and are assumed to have had an impact as far westwards as the River Rhine valley.

The Nordic Ice Sheet, which advanced again into Central Europe during the subsequent Weichselian Ice Age, had a much smaller extent: the maximum ice margin ran north–south through the central parts of Schleswig–Holstein and did not cross the River Elbe valley (Hughes et al. 2016). Since these forms are much younger and were not subsequently modified by periglacial processes, a great diversity of field evidence remains. As the ice did not advance as far as the low mountain ranges and spread throughout the lowlands, no major ice-dammed lakes were generated. With regard to drainage dynamics, however, the ice-marginal glacial valleys (german "Urstromtal") deserve special attention (Liedtke 1981). These formed during the melting down of the ice shield after the peak of the last ice age about 20,000 years ago. This statement can be made because they are not covered by moraine deposits at any point, which is why the active ice margin must have displaced already to the north of them. They formed during the stadials of the

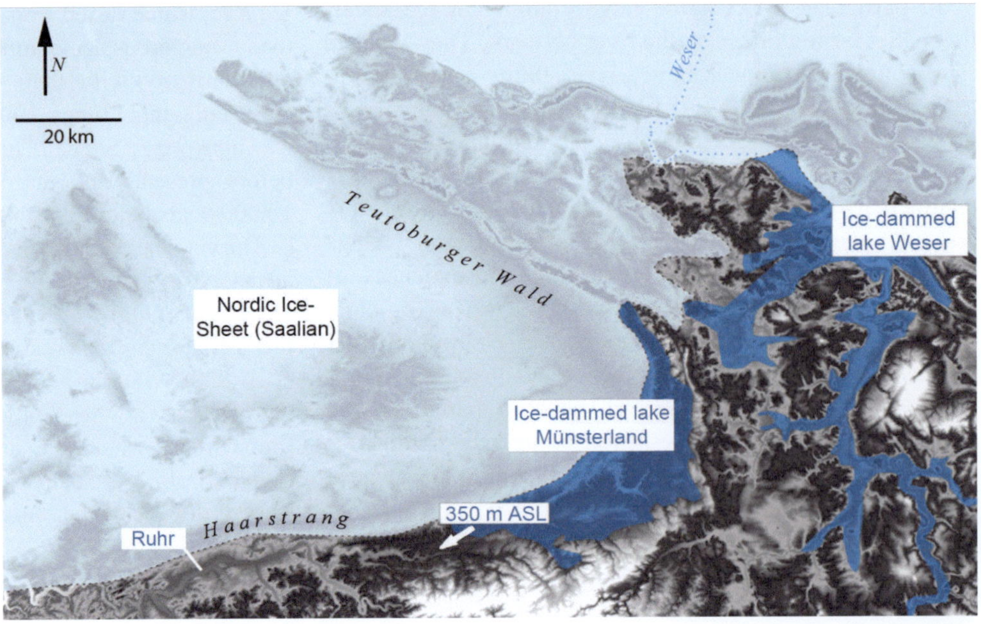

Fig. 5.1 The ice-dammed lakes in the River Weser valley and in the south-eastern part of the Münsterland Embayment (western Germany) during the Saalian Ice Age (modified after Meinsen et al. 2011)

Weichselian Ice Age, during which the retreat of the active ice margin paused and a stable ice margin developed over a period of time. The rivers flowing from southern directions against the ice margin were blocked and, together with the meltwater draining from the ice, discharged parallel to the ice margin following the gradient to the west. As conditions were in a stable position for several centuries, these ice-marginal glacial stream valleys were able to incise. By this process pattern, a corresponding glacial valley developed parallel to each stationary ice margin. They form an extensive network through the North German Lowlands east of the River Elbe valley. Only locally flood event related features could be identified so far (e.g. Weckwerth et al. 2019).

5.1.1 The Baltic Sea—A Post Glacial Ocean Basin

At the end of the last ice age, the active ice margin was displaced back towards Scandinavia, exposing the basin of the Baltic Sea again (Hughes et al. 2016). Precipitation and meltwater accumulated along the depth contours and began to form a lake, which subsequently rose rapidly because the basin had no drainage outlet. Around 11,700 years ago, during the so-called Younger Dryas Period, the active ice margin was located in central Sweden and the water level of the lake had risen so high that the southern ice margin was acting as a dam. However, the sea level of the Northern Sea was still much lower than today (Vink et al. 2007), because a lot of water was still stored in the ice and thus missing from the ocean basins. Therefore, southern Sweden was connected to Denmark via a land bridge, so the Öresund was still closed (Björck 1995). The lake in the Baltic Sea rose up to a height of 25 m above sea level, which, due to the topographical conditions at Mount Billingen in central Sweden between today's Lake Vänern and Lake Vätter, developed into an important key point. Field evidence by lake deposits and shorelines reveals that the basin of the highly risen ice reservoir of the Baltic Sea and another basin in extension of the sea depth of the Skagerrak almost touched each other here. If one tries to imagine the topography of that time today, one must take into account that, on the one hand, the sea level was significantly lower, but on the other hand, Scandinavia had also sunk into the earth's crust as a result of the imposed load of the inland ice and the compensating isostatic uplift had not yet taken place completely (Ehlers 2022). Therefore, today's topography differs significantly from the situation at that time (Fig. 5.2).

The importance of the locality around Mount Billingen has long been known to researchers, but controversial views on the emptying of the Baltic Ice Reservoir behind the ice barrier have long been debated (Björck 1995; Brunnberg 1995; Johnson et al. 2022). Only with the discovery of obviously fluvial transported blocks southwest of the mountain, the dating of shore deposits of the Baltic Ice Lake, and the identification of submarine landslides at the bottom of the present-day Baltic Sea (Hyttinen et al. 2011) did it become clear beyond doubt that there must have been a rapid lowering of the water level and thus an outburst flood from the Baltic Ice Lake occurred (Fig. 5.3).

Jakobsson et al. (2007) have calculated that the outburst of the Baltic Ice Lake released about 7800 km^3 of water,

Fig. 5.2 The location of Mount Billingen and extent of the Baltic Ice Lake about 11,700 years ago. The data on the ice margins, sea level height and extent of the Baltic Ice Lake are taken from Björck (1995) and are schematised on a satellite image montage (NASA Visible Earth). Areas where the present-day surface representation emerges from the satellite image were land-locked, be it present-day mainland as in southern Sweden or present-day seabed of the North and Baltic Seas. Due to glacial isostatic land subsidence from ice loading and eustatic sea level subsidence from water storage in the ice, the land-sea distribution is different from today

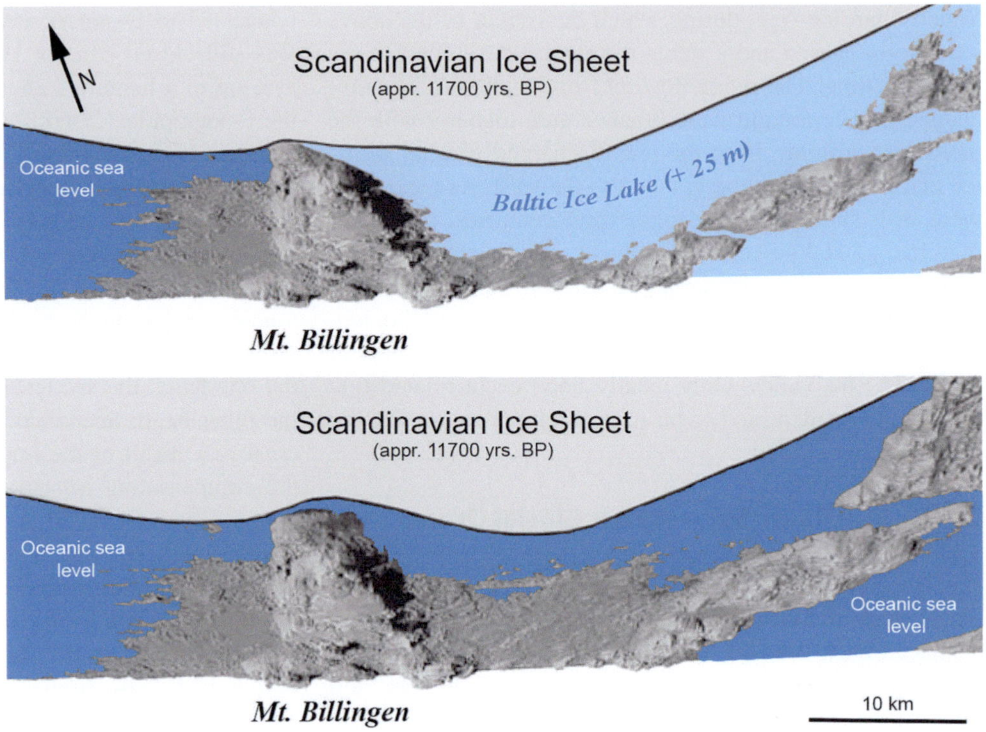

Fig. 5.3 Elevation model of the outburst site of the Baltic Ice Lake at Mount Billingen (modified after Jakobsson et al. 2007). After the outburst of the Baltic Ice Lake, the water level declined by around 25 m. The basin remained connected to the North Sea, so salt water could flow into the area

which corresponded to about 18% of the total volume. It is unclear how long it took for the water level to level on both sides. Assuming that 1–2 years were needed for this, the inflow of fresh water into the North Sea was 120,000–250,000 m^3/s. Subsequently, the interaction between sea-level rise, the inflow of fresh water from Central Europe and the melting inland ice over Scandinavia, as well as the glacial isostatic rise of Scandinavia, continued. Inland seas with salt, brackish and fresh water developed through repeated opening and closing of the connection of today's Baltic Sea with the North Sea (Lampe 2002), until finally, about 8000 years ago, the present situation of the connection of the North Sea and the Baltic Sea stabilised.

5.1.2 Britain Became an Island—Formation of the English Channel

Pleistocene sea level changes also played a major role for the British Isles, because when sea level was low, a land bridge was formed and the archipelago became part of mainland Europe (Preece 1995). The question of the origin of the English Channel has a correspondingly long tradition of research. In this context, considerations of a catastrophic breach caused by the erosive force of flowing water appeared early on Smith (1985), Gibbard and Lautridou (2003), Gupta et al. (2007), and Garcia-Moreno et al. (2019). However, the various concepts had the problem of providing robust evidence and conclusions on the origin of the water and the timing of the destruction of the land bridge. It has long been known from bathymetric surveys that the floor of the English Channel consists of a network of valley channels. Within the framework of high-resolution detailed images (Fig. 5.4), Gupta et al. (2007) were able to reveal that the valley channels can be divided into several valley generations, i.e. they have different ages.

Based on the recordings of the topography of the seabed in the area of the English Channel, it can be concluded that until the Elsterian Ice Age, a permanent land bridge of solid rock existed between the European Mainland and the British Isles with a core slightly west of the Strait of Dover. During the Elsterian Ice Age, the Nordic Ice Sheet over Scandinavia and the ice sheet on the British Isles were connected. This prevented the rivers draining the mainland, such as the Rivers Rhine and Meuse, from flowing north into the North Sea and dammed up a lake in the southern North Sea. Its rising water level eventually reached the elevation of the land bridge and the water began to pour into the valleys west of it. As these valleys were already incised to some degree, there was a relatively steep gradient on the west side of the devide, which meant that the flow velocity of the spilling water was high, allowing it to incise further on and cut backwards quickly. In this process, the spilling water continuously increased, so that a corresponding outburst flood could form another valley generation and destroy the land bridge. For the estimation of flows during the Elsterian period outburst flood, Gupta et al. (2007) were able to draw on detailed information on channel geometry

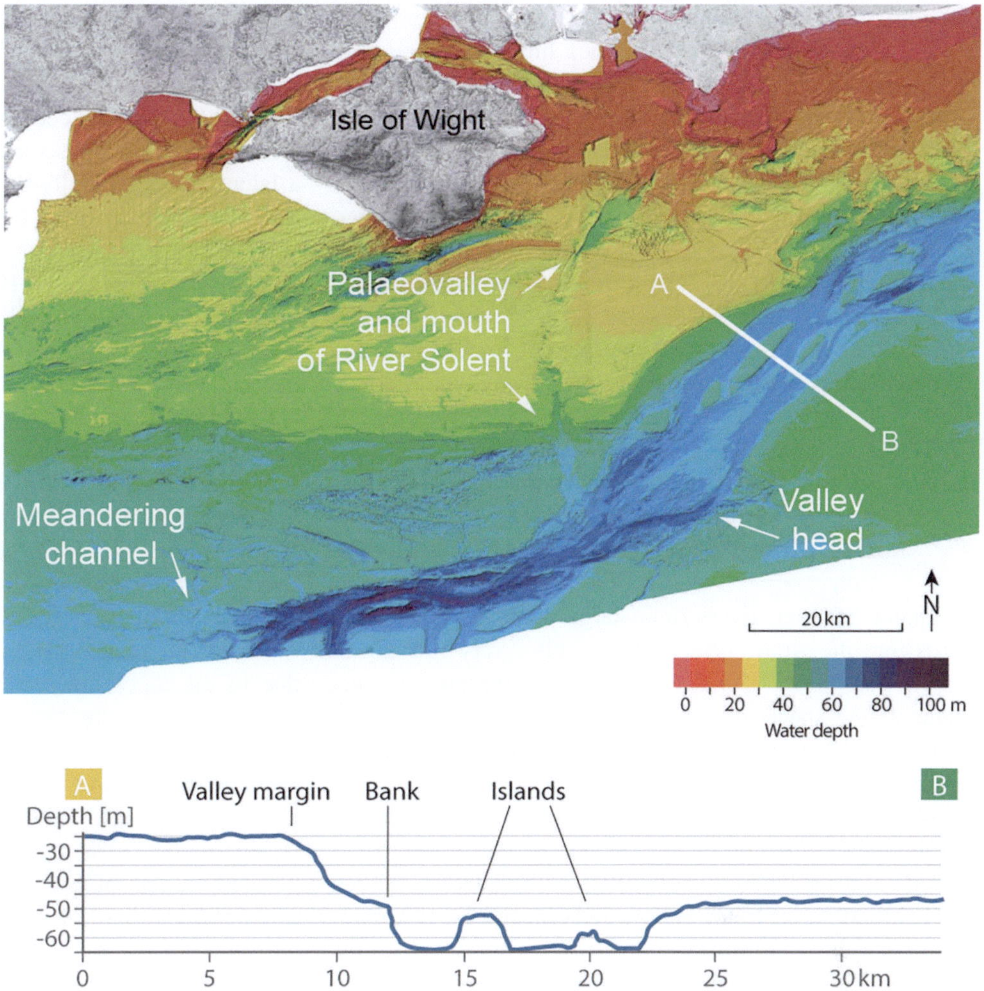

Fig. 5.4 Bathymetric map and profile showing parts of the English Channel with evidence of valley systems of different ages (modified after Gupta et al. (2007) from Herget 2010)

such as width, depth and gradient, but had no information on water levels within the channels. Under the simplifying assumption that the individual channels were filled to the top, i.e. that there was bankfull discharge, discharges of 200,000–1,000,000 m^3/s could be estimated. So it was an ice dam outburst that gave Britain its island status. This was at least during the warm periods when sea levels were high. During the subsequent glaciation periods of the Saalian and Weichselian ice ages, the sea level lowered again, but then a large river fed by melt and precipitation water flowed through the present-day channel, the so-called Channel River or in French "Fleuve Manche". Toucanne et al. (2009, 2010) correlate its water flow with the extent and position of the ice sheets in Europe. The background is the consideration that, coherent ice sheets from the British Isles and Scandinavia in the North Sea block the outflow to the north and so all water must run through the Channel River into the mid-Atlantic. The inflow of freshwater into the northern Atlantic was thus reduced, which may have had an influence on the thermohaline circulation (cf. below Sect. 5.3.1 on Lake Agassiz). But also the development of the Neanderthals' habitat and the animal herds they hunted was undoubtedly lastingly influenced by the water network with the large meltwater streams that could hardly be crossed (Hijma et al. 2012). Repeated archaeological findings confirm, that the seabed of nowadays southern North Sea was inhabited by man (e.g. Tizzard et al. 2014; Davis et al. 2023).

5.2 Siberia—Less Ice but Large Lakes in the Cold Desert

In Pleistocene times, also Siberia with its characteristic continental climate experienced the global cooling pattern and expanded glaciation. The chronostratigraphical pattern is similar and mostly synchronous to the rest of the world (e.g. Cohen and Gibbard 2022). The situation is getting more complicated by focussing on the glaciations, especially their volume respectively extension (cf. e.g. discussions in Margold et al. 2016, Batbaartar et al. 2018 and Nørgaard et al. 2023). Due to the continental climate, moisture supply to continental Central and Northern Asia is (and was during the Pleistocene) limited. As glaciers require both low temperatures and sufficient moisture supply to form and expand, the glaciation dynamics in the region were not necessarily synchronous with the global pattern. Consequently, terms like "last glacial maximum (LGM)" should be used with caution or even better avoided as interpretations and conclusions could be easily misleading. There are some hints about this from previous studies (cf. Herget et al. 2020 for references) which are also relevant for the investigation and interpretation of Pleistocene ice-dammed lakes and related outburst floods in the region.

5.2.1 Forced Drainage Towards South—Ice Shields Blocked Drainage in Continental Scales

All major Siberian rivers drain large parts of Asia northwards into the Arctic Ocean, most with a very low gradient. As part of the global cooling during the last ice age, extensive ice sheets developed here as well, blocking the mouths of the rivers and thus leading to the formation of very extensive ice-dammed lakes (Fig. 5.5).

Around 90,000 years ago, the ice sheet over Scandinavia and the adjacent Arctic Ocean reached its maximum extent towards east (Mangerud et al. 2004). Subsequently, the ice sheet shifted its centre of gravity and reached its maximum extent in Central Europe about 20,000 years ago. At this time, the ice extent was again reduced from the Siberian mainland to the area of the Kara Sea. During the maximum expansion in northern Russia and Siberia, rivers such as the Dwina, Petschora, Ob and Jenissej, which flowed into the glaciated area, were back up dammed over almost 2000 km and formed extensive lakes. Three independent lakes were generated in the lowlands with water levels at 100 m (White Sea Lake—WM), 90–110 m (Komi Lake—K) and 60 m (West Siberian Lake—WS) above sea level, which had a total volume of 32,200,000 km^3. There are considerations of outburst floods from the lakes that could be expected when the active ice margin receded and the drainage reopened. Mangerud et al. (2004) estimate peak discharges of 300,000–850,000 m^3/s with a duration of 2–3 months for Lake Komi. Knies and Vogt (2003) analyse evidence from boreholes in the floor of the northern Barents Sea, which document distinctive freshwater inflows, among others, around 77,000 years ago. But these are not the main characteristics of these Pleistocene ice-dammed lakes. What is more important here is that the lakes reached water level elevations that surmount the continental watershed divide.

The ice-dammed lake above the White Sea drained into the Baltic Sea through spillways in the area of Lake Onega and Lake Ladoga. Due to the already lower sea level, the Baltic Sea was at this time an inland sea cut off from the world sea. It was subsequently completely covered by the advancing Scandinavian Ice Sheet and developed into the Baltic Ice Lake (cf. Sect. 5.1). The White Sea Lake was also influenced by the subsequent change in the ice margins and a later developed spillway into the watershed of the River Volga and thus into the Caspian Sea. The Caspian Sea, as an inland sea, had a much higher water level than today and thus a greater extent. Based on lake deposits and shorelines, the Caspian Sea had a water level of 26 m above sea level. The water level could not have risen higher than this, because a spillway through the Manych Depression (which is dry today) came into operation and transferred the water via the Sea of Azov into the Black Sea. Activity of the spillway from the Caspian to the Black Sea has been proven for this period by dating (after Leonov in Mangerud et al. 2004), which provides evidence of the drainage of Eastern Europe into the Black Sea not only via the Rivers Dnipro and Don (Sidorchuk et al. 2011), but also via the Volga.

While there are no striking spillways for Lake Komi due to the topography, this is different for the largest of the lakes behind the Urals. Here, around 90,000 years ago in the early stages of the Weichselian Glaciation, the large West Siberian Rivers Irtysh, Ob and Yenisey dammed up Lake West Siberia. With a volume of 15,000 km^3, it contained almost 50% of the total volume of all lakes and covered an area of 610,000 km^2. With a lake level of 60 m a.s.l., today's divide to the Aral Sea could be crossed via the valley of the River Tobol and through the Turgai Basin in the northwest of present-day Kazakhstan. Its present elevation of 126 m is due to younger deposit accumulation (Mangerud et al. 2004) and the bedrock surface below it is at an elevation of 40 m a.s.l.. About 90,000 years ago, the divide was probably located further north, where a sill in the bedrock surface of 55 m height is identified. However, when Lake Aral was reached, which at that time had a higher water level of around 57 m a.s.l. and a correspondingly greater extent, the cascade of spillways continued in south-western direction. Via the Uzboy Pass, the Caspian Sea, located in a tectonic depression, was reached, which also had a higher water level. Its maximum water level, which today is below that of the World Sea, was limited by the Manych Depression as a connection to the Black Sea. Due to the backwater in the large lowland valleys of the West Siberian streams, the drainage of the continent was temporarily reversed: instead of Kara Sea in the north, the Black Sea is the south was reached via intermediate steps. Since the connection of the Black Sea to the Mediterranean Sea was interrupted by the globally lower sea level, the water accumulated here and evaporated. Similar to today, the Black Sea also had an overall negative water balance during the ice ages due to high evaporation and thus a significantly lower water level. The replenishment to today's

Fig. 5.5 Ice-dammed lakes in northern Eurasia and their spillways about 90,000 years ago and 20,000 years ago (modified after Mangerud et al. 2001 and 2004)

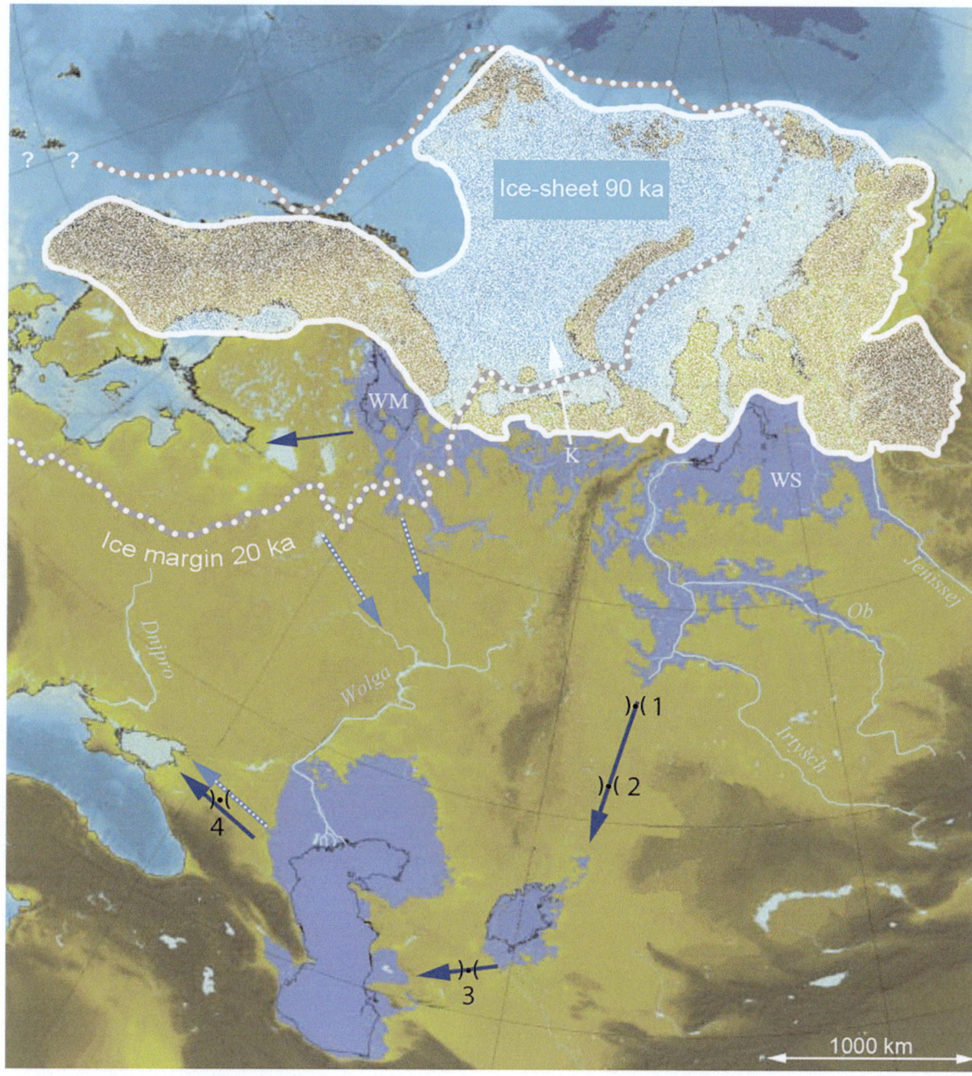

Ice-dammed lakes:

WM	White Sea
K	Komi
WS	West-Siberia

Spillways:

90000 years ago

20000 years ago

Spillways:

1 Ridge bedrock 55 m
2 Turgai Basin
126 m surface
40 m bedrock
3 Uzboy Pass 57 m
4 Manych 26 m

water level corresponding to the Mediterranean Sea had potentially dramatic features (cf. Chap. 1).

Russian scientists have long been investigating the extent of the Pleistocene glaciations in Siberia and the resulting consequences for the water network (e.g. Astakhov 1991; Arkhipov et al. 1995) and have developed different hypotheses. For example, the supply of the Central European ice-marginal valleys via spillways from Siberian ice-dammed lakes or outburst floods that poured over thousands of kilometres across the West Siberian Plain and the steppes of Kazakhstan (Grosswald and Rudoy 1996; Grosswald 1999; Grosswald and Hughes 2002). The dating techniques now available prove that the chronology of the change in the extent of the Nordic Ice Sheet hindered a connection of the northeast European ice-dammed lake with the ice-marginal valleys of Central Europe due to missing synchronicity (Fig. 5.5). Nor is there any plausibility for interpreting the parallel linear structures in the plain west of Siberian city of Barnaul (Fig. 5.6) as erosion channels of outburst floods. For this, the region at the headwaters of the River Ob lacks

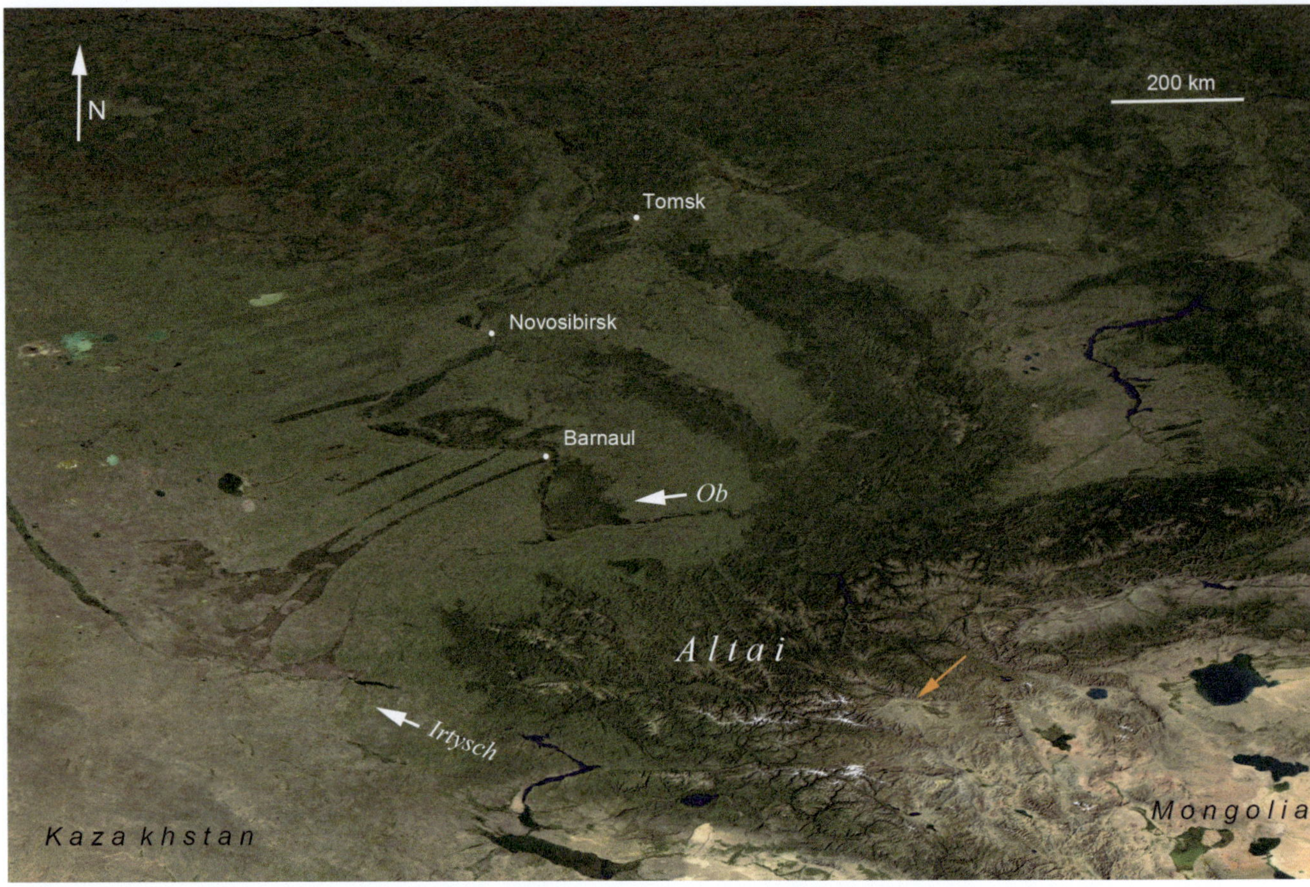

Fig. 5.6 Location of the Russian Altai Mountains in southern Siberia (satellite image from NASA Visible Earth). The prominent linear shapes west of Novosibirsk and Barnaul are tectonic structures rather than erosional channels of glacial outburst floods. The yellow arrow marks the location of the Chuya Basin

source of the required masses of water and the driving force to move southwestwards across the extensive plain against the basic northwestern gradient. The linear forms strongly emphasised by the different vegetation cover are probably rather tectonic structures.

5.2.2 Repeated Outburst Floods in the Russian Altai Mountains

In addition to ice-dammed lakes at the margin of the Nordic Ice Sheet, there have also been local mountain glaciations in Siberia where ice reservoirs have formed, producing catastrophic outburst floods (e.g. Komatsu et al. 2009; Batbaartar and Gillespie 2016a, b; Margold et al. 2018). As a relatively well-studied example—partly controversial reviews are offered by Baker et al. (1993), Butwilovsky (1993), Rudoy (2002), Carling et al. (2002), Herget (2005), Zolnikov et al. (2015), Krivonogov et al. (2017), Bohorquez et al. (2019), Herget et al. (2020), Agatova et al. (2020)—the repeated ice dam outburst floods in the Russian Altai Mountains (Fig. 5.6) are presented in more detail.

In the up to 4500 m high mountains in southern Siberia, where the headwaters of the Bija, Katun and Chuya Rivers as major sources of the River Ob are located, a typical mountain valley glaciation developed during the ice ages. The advancing glaciers blocked the course of the River Chuya, thus damming up an ice-dammed lake (Fig. 5.7).

The ice dam consisted of five valley glaciers that flowed into the Chuya valley from different directions, forming an ice dam in the area of the village of Aktash about 30 km long. Considering the altitude of the valley floor and the maximum elevation of shorelines (Fig. 5.8) remaining of the dammed lake at 2100 m a.s.l., the glaciers must have been at least 650 m thick. With a surface area of 2600 km^2 and a volume of 607 km^3, the over 100 km long ice-dammed lake filled the entire Basins of Kuray and Chuya.

In addition to the shorelines, there is also further evidence of the lake (Carling et al. 2011 for further details). In the valley of the Chagan-Uzun River in western Chuya Basin, extended varve deposits from an earlier lake stage

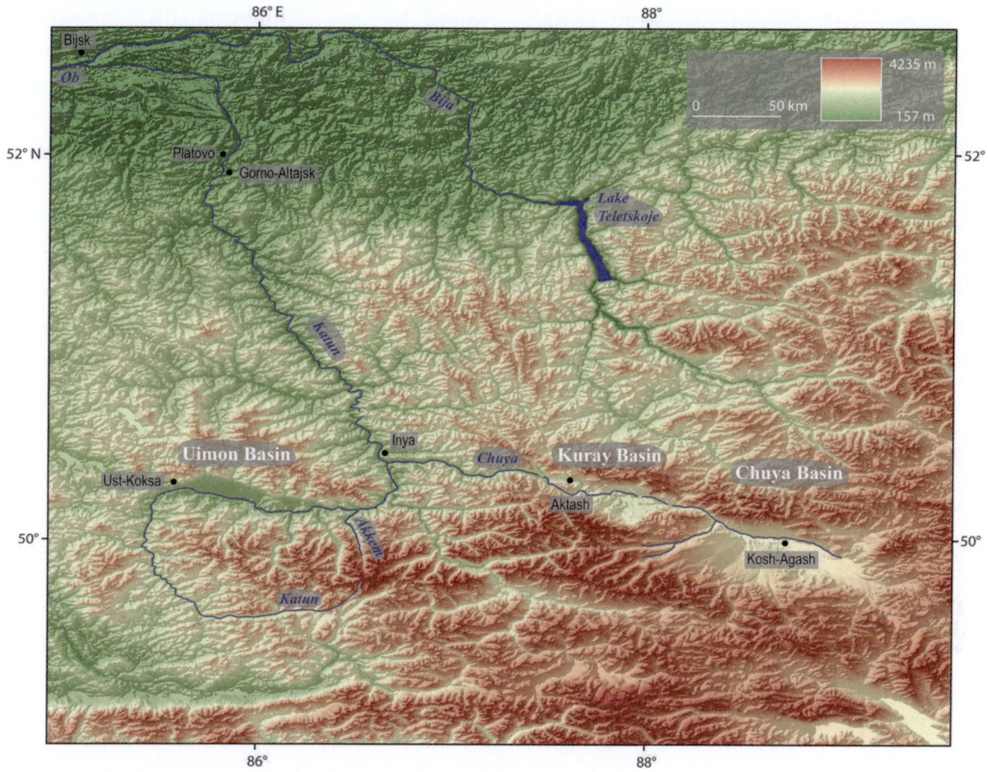

Fig. 5.7 Review map of the catchment areas of the Rivers Bija, Katun and Chuya, sources of the River Ob in the Russian Altai Mountains (modified from Herget et al. 2020, based on data from ASTER GDEM 2009)

Fig. 5.8 Remaining shorelines in Kuray Basin (persons for scale). The shorelines are generated by the erosive effect of wave action on the shore and therefore indicate the elevation of the lake level. The shorelines visible here originate from a low intermediate stage of the lake, while the highest ones can be found at 2100 m altitude (Carling et al. 2011 for further details)

lie in undisturbed bedding. The remaining lake sediments in other locations within the basin have been redeposited by the high-energy currents during the drainage of the lake, so that the characteristic sediments can still be found, but not their original bedding structure (cf. map in Herget et al. 2020 for the locations). The currents within the lake have created beach spits and tombolas, which are well preserved in the eastern part of Chuya Basin. There are also numerous drift blocks that were initially carried by ice floes, floating on the lake surface, but then melted out or stranded and today form eye-catching deposits as erratic blocks differing by their lithology from local bedrock (Fig. 5.9).

Downstream the former ice dam, the giant bar deposits along Chuya and Katun valleys bear witness to the water level of the outburst flood waves (Fig. 5.10) (Carling et al. 2009; Carling 2013 for further details). The gravel bars consist of fine gravel transported in suspension, and were deposited at local valley widenings at the mouths of side valleys as still water sediments. The elevation of their surface indicates the minimum water level of the outburst flood waves as a palaeostage indicator (cf. Chap. 2.2 on palaeostage indicators). After the passage of flood waves, the entire valley floor was filled to a depth of about 50 m with significantly coarser sediments. As indicated by foreset structures in local outcrops, these sediments were transported in a rolling and sliding mode on the valley floor and is therefore referred to as a bedload terrace.

While the fine-grained suspension-load deposits of the giant bars had not filled the valley to its full width and had

Fig. 5.9 Drift block deposit in the north-eastern part of Chuya Basin at 2004 m altitude

Fig. 5.10 Gravel bars as palaeostage indicators in Katun River valley at the village of Inya. The boulder field in the foreground is located on the bedload terrace that fills the entire valley bottom

thus temporarily raised the valley floor, the deposits cannot be addressed as a terrace. Terrace formation can be ruled out because the grain size of the deposits with a typical diameter of < 1 cm (Fig. 5.11) does not come to deposition even at today's comparatively low flow velocity and can therefore only be explained as slack water deposit from suspension during an outburst flood.

The deposition of suspension gravel has blocked the mouths of numerous tributary valleys along Chuya and Katun River and thus frequently led to backwater lakes therein. The typical fine-grained, yellowish lake sediments were deposited in these lakes until their water level reached the bar's surface and the lake drained through the resulting spillway. In one of these tributary valleys near Inya village, there is now a twofold repeated overburden of these lake sediments with suspension gravels (Fig. 5.12), indicating that there must have been at least three large outburst flood wave that flowed through Katun River valley.

The dynamics to be inferred from the alternating sequence of lake deposits and suspension gravels in the tributary valley can be imagined as follows: A first ice-dammed lake outburst flood causes the deposition of suspension gravels in the tributary valleys and a blockage of their outlets by the deposition of gravel bars (Fig. 5.10). Backwater of runoff in the tributary valleys results in deposition of fine-grained lake sediments until the lake forms a spillway in the gravel bar and drains from the tributary valley into Katun River valley. This breach remains open until another outburst flood flows through Chuya and Katun valleys many decades to centuries later, depositing its suspension gravels through the breach back into the tributary valley. Here, the lake deposits are covered by suspension gravels and the breach, which is a still water area for the main wave through the valleys, is closed again. Once again, the water in the tributary valley backs up and deposits fine-grained lake sediments on the suspension gravels. As the bedding conditions in the tributary valley near Inya document, there have been at least 3 outburst flood waves of similar high magnitude. Sequence analysis of the sedimentary structure of the giant bars confirms the indicated occurrence of repeated outburst floods by an even higher number events (cf. Carling et al. 2009 for details).

Fig. 5.11 Section illustrating the characteristic fine-grained suspension gravel within the giant bar at Inya village (Prof. Paul Carling for scale). More vivid than a grain size analysis, the figure shows the characteristic fineness of the outburst flood deposits. Such kind of deposit is characterizing a megaflood deposit according to Carling and Fan (2020)

The outflowing water has also left behind striking forms and structures in the lake basin. For example, fluvial gravel dunes can be found in the valley of the River Chagan-Uzun on the western part of Chuya Basin (Fig. 5.13; cf. also Fig. 2.30) among other locations. Note the dimensions, as the fluvial gravel dunes at Chagan-Uzun valley have heights of up to 23 m and wave lengths of 320 m (Herget 2005).

Furthermore, scour structures on solid rock mounds have also been preserved (Fig. 5.14). When flowing out, the water at the bottom of the lake had to flow around an obstacle in the form of a rock mound. This resulted in increased flow velocities in front of and at the edge of the mound, so that sediments lying there were eroded to form a scour (cf. Chap. 2.3 on obstacle marks). The scour itself indicated by the green grasses is 9 m deep, 340 m wide and 92 m long. Unfortunately, the reworked sediments once deposited in the wake behind the hill has since been eroded again by River Chuya.

The outburst flood waves moving down along Chuya valley also rose against the gradient of the Katun valley upstream into Uimon Basin (Fig. 5.7) according to the elevation of the water level indicated by the giant bar surfaces. During the continuation of the flood wave movement, the water came to a standstill in Uimon Basin and the mouth of the Katun was blocked by the suspension gravel deposits (Fig. 5.15). After the outburst flood had subsided, the water backing up in the upper Katun valley and Uimon Basin eroded a breakthrough breach through the deposits. However, as illustrated by the spatial arrangement of the gravel bars at the River Chuya mouth into Katun River and the narrowly cut channel of the Katun, even subsequently

Fig. 5.12 Alternating sequence of lake deposits and suspension gravels in Inyushka River valley, a tributary valley near Inya village. Visible is the uppermost of a total of three sequences of grey suspension gravel (arrows at lower end within three ridges separated by recent gullies) embedded between light lake sediments

Fig. 5.13 Fluvial gravel dunes in the valley of the Chagan-Uzun River, westerm Chuya Basin (flow direction to the lower left)

the much larger river has not been able to clear out these deposits again (Fig. 5.15).

This interpretation is supported by evidence of indication of bidirectional flow in deposits in the Katun valley between the mouth of Chuya River and Uimon Basin (Carling et al. 2010). The backwater in the Uimon Basin can also explain lake deposits there, to which no ice dam can be attributed, because the maximum ice margin elevations were much higher up in the mountains at the time of the outburst flood waves (cf. Rudoy 2002). Based on hydraulic modelling for unsteady flow conditions (Bohorquez et al. 2019) the inundation of Uimon Basin by the flood wave from Chuya River valley lasted for about 2 days only.

Temporally, the outburst flood waves can be narrowed down to the period of Marine Isotope Stage MIS 2 between 29,000 and 15,000 years ago by focussing on the most prominent remaining features (cf. Herget et al. 2020 for literature review and systematic discussion). The wide range can be explained by the small amount of datable material available and the use of different dating methods. Difficulties arise, for

Fig. 5.14 Scour structure formed by the draining waters within the Pleistocene lake near Chagan-Uzun village, western Chuya Basin

Fig. 5.15 The spatial arrangement of gravel bars in the area of the confluence of Chuya and Katun Rivers indicates dominant flow out of Chuya Valley. Note the winding course of Katun River as its valley is remained blocked by bars consisting of fine-grained suspension gravel

example, from the fact that in the semi-arid high mountain region little organic material is available for the application of the radiocarbon method. Furthermore, many rocks are too impure to obtain sufficient material for exposure ages. For luminescence techniques, a problem of insufficient bleaching due to the short and turbulent transport in the outburst flood wave was revealed. Recently, several dates of flood related samples were published (cf. Herget et al. 2020) that derived ages of 80–90 ka, fitting more or less to MIS 4. In previous investigations, such ages had to be put in perspective due to indicators of limited bleaching. Consequently, such ages require verification both analytically and in the environmental context (e.g. glacier extensions, post-flood periglacial reworking, …). Note, that previous additional outburst floods should be expected and are plausible in the context of previous glaciations, which are still under discussion and therefore remain as an open question.

The outburst floods in the Russian Altai Mountains served as an example for the development and testing of different approaches to the reconstruction of their discharge, so that a large number of data on the peak discharge, but also various intermediate stages, are available (Baker et al. 1993; Herget 2005; Carling et al. 2010; Bohorquez et al. 2015; Agatova et al. 2020). For the peak discharge, values of 10,000,000 m^3/s with flow velocities of typically 20–30 m/s are determined and confirmed according to a variety of methods. Important for this estimation are the surface elevations of the gravel bars of suspension gravels, which serve as palaeostage indicators at numerous locations along Chuya and Katun River valley. They also document the maximum water depth of up to 400 m in Chuya Valley and around 250 m in the much wider Katun Valley (Fig. 5.16).

These discharge values can also be confirmed by the run-up sediments. For this approach based and developed on the field evidence in the Russian Altai Mountains (Herget 2005), there is thus also the possibility of plausibility testing of the estimates derived from other modelling approaches of flow velocity. The suspension gravels deposited in front of valley constrictions such as rock ridges, due to water level rise towards higher elevations by energy transfer from kinetic into potential energy (cf. Chap. 2.3), are partly more than a hundred metres higher than the nearest gravel bar (Fig. 5.17).

It is problematic that, on closer inspection, the surface of the undisturbed comparative gravel bar only indicates a minimum water level, as the suspension load from any height down has accumulated there. Therefore, the height

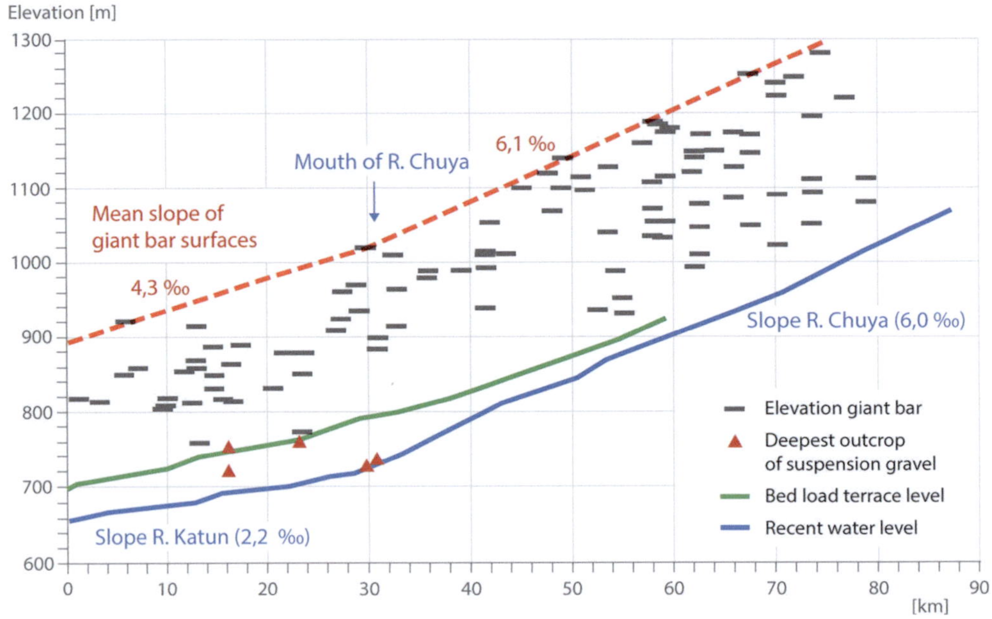

Fig. 5.16 Elevation of the surfaces and terraced levels of the giant bars along Chuya and Katun valleys (modified from Herget 2005, a data table is available therein)

Fig. 5.17 Run-up sediments in the lower Chuya valley (flow direction to the left). The difference in elevation to the surface of the gravel bar is 124 m; from a different perspective, the same location is presented in Fig. 2.39

difference cannot be estimated with certainty. On the other hand, the valley has a trapezoidal profile, so it becomes wider towards the top. If one now assumes a higher water level than the immediate gravel bar surface, the difference in elevation to the highest-lying run-up sediments and thus the estimated proportional flow velocity v becomes smaller. At the same time, however, the cross-section area A increases. This increase of A is compensated by the decrease of v in the calculation of the discharge Q as a product of both, so that in practice the uncertainty in the estimation of the former water depth does not play a significant role (cf. Herget 2005 for details, Carling et al. 2020 for additional theoretical comments). In this assessment, it must be taken into account that we are dealing with a plausible estimation of magnitudes of discharge in the order of millions of cubic metres, so that differences respectively errors of 100,000 m^3/s can already be considered as non-significant values in the decimal range.

Based on a known peak discharge and the flood volume limited by the lake size, the duration of the passage of the outburst flood wave can easily be estimated based on the flood's hydrograph (cf. Sect. 2.4). For a peak discharge of 10 million m^3 and assuming the emptying of the entire lake volume of 607 km^3, this results in about 2.5 days (Fig. 2.41). This value does not take into account the speed of the opening of the breach in the dam, nor does it take into account any retention effects, for example in side valleys along the Chuya and Katun valley courses. Modelling the dam breach for unsteady flow yields values of a comparable magnitude (Carling et al. 2010; Bohorquez et al. 2015, 2019). None of these estimates of flood duration should not be taken literally, but the statement that the flood duration lasted less than a week can be made with certainty.

Despite the enormous dimensions of the repeated outburst flood waves when they were channelled through the valleys of the Russian Altai Mountains, they are comparatively small in relation to the vastness of Siberia (Fig. 5.6). Considerations that this outburst flood wave had a noticeable impact as far as the Arctic Ocean or contributed as an important source of water to the overflow into the Black Sea (Grosswald and Rudoy 1996; Reuther et al. 2006) can therefore probably be viewed rather critically.

Anyhow, there are numerous flood related features that are worth a closer look, not only for their impressive dimensions, but also for the necessity to interpret them in individual context. The deposition of boulders (Herget 2005), the formation of scours around them afterwards (Herget et al. 2013) or the formation of channel bedforms such as fluvial gravel dunes (Carling 1996; Herget 2005) is more likely to be associated with an unspecific stage during the decay of the flood. It must also be taken into account that some structures were created in the lake basin, so they cannot provide any information about peak discharges of the outburst flood downstream.

5.3 Pleistocene Outburst Floods in and from North America

The Laurentian Ice Sheet, the valley glacier systems of the Rocky Mountains, and even local alluvial fans formed dams and subsequent outburst floods in North America, which led to a fundamental reshaping of the relief and also triggered global climate fluctuations. Three examples illustrate the range of dynamics during in North America during the last Pleistocene glaciation.

5.3.1 Continental Scale Ice-Dam Lake Dynamics—Lake Agassiz

One has to think on a larger, literally continental scale when considering the formation and dynamics of ice-dammed lakes at the margin of the Laurentian Ice Sheet in North America during the waning of the last ice age. The ice sheet had its centre in northern Canada and retreated towards there during the decay. About 9000 years ago, the centre was located over the Hudson Bay area (Fig. 5.18).

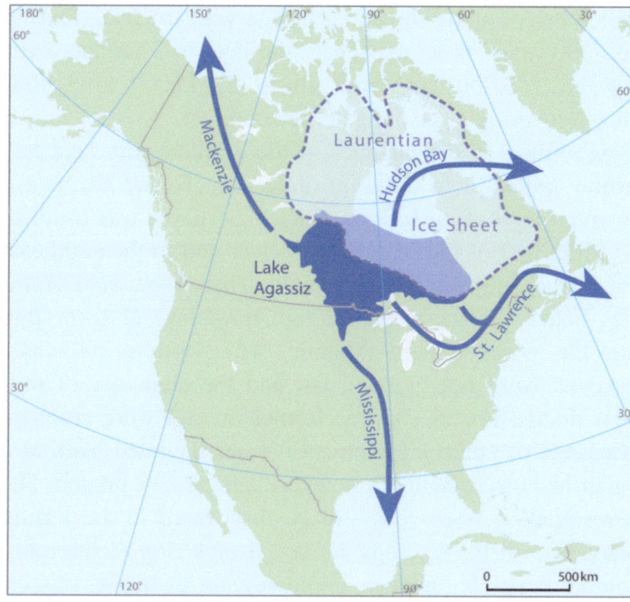

Fig. 5.18 The Laurentian Ice Sheet over the Hudson Bay with Lake Agassiz (modified after Teller et al. 2002)

The ice sheet blocked the drainage of the continent by the numerous rivers flowing into the Hudson Bay, and an extensive ice-dammed lake formed in front of the ice. Corresponding to the shifts of the lobes of the active ice sheet, the thawing of dead ice remnants and the melting of ice remnants in moraine walls, a constantly changing picture of lakes and spillways formed during the end of the glaciation (Lewis and Teller 2007; Kehew et al. 2009). The situation shown in Fig. 5.18 reflects the basic situation at the end of the last glaciation respectively in early Holocene times, when the lake named after the glaciologist Louis Agassiz had reached its maximum extent.

The reconstruction of the ice-dammed lake, its spillways and outburst floods is complicated by the glacial isostatic uplift of the North American continent. As a glance at an atlas map illustrates, large parts of eastern Canada are located at an elevation of less than 500 m above sea level. This is also the amount of glacial isostatic uplift due to the reduction of the stress by the ice sheet. At such amounts, the watercourse networks and flow directions of the rivers draining the continent took time to develop into the present-day pattern and flow directions could also be reversed. Therefore, no watercourse network is shown in Fig. 5.18. The same applies to the shorelines resulting from wave action, which trace the extent and elevation of the ice-dammed lakes. Today, they show slopes of partly more than 200 m over a length of 500 km (Teller 2001), which illustrates the dimensions of glacial isostatic crustal deformation in North America. Due to the permanent changes of the lake and its sub-basins, it is difficult to make representative size statements, which is also due to the uncertainties regarding the ice extent along the contact zone with the lake. During the 4000 years of its existence, the lake had surface areas between 150,000 km^2 and almost 1,000,000 km^2, thus always covering large parts of the continent.

According to the changes in the gradient, the lake had several spillways in different directions. To the northwest, the river network of today's Mackenzie River was used, to the south that of the Mississippi River and to the southeast that of the Saint Lawrence Stream. There were also drainages below the ice sheet through nowadays Hudson Bay into the north-western Atlantic. The various spillways, however, were not gradual, but had the character of outburst floods. For as soon as former divides were reached, thousands of cubic kilometres of water dammed back at a higher level were ready to flow through the new breach. The corresponding traces can still be discovered in the terrain today, even if the findings are not always easy to interpret. Boulder deposits found in a spillway, for example, suggest discharges of 102,000–364,000 m^3/s (Fisher 2004). The traces of these outburst flood waves through the Mississippi valley can be traced down to the sea floor in the Gulf of Mexico and thus allow a comparative classification in the discharge dynamics of the great river during the last ice age as a whole. It can be shown that this spillway from Lake Agassiz was only used until about 13,000 years ago, after which drainage predominated due to the change in gradient caused by isostatic uplift in other directions (Tripsanas et al. 2007). For events with drainage into the Arctic Ocean, a discharge of up to 2,500,000 m^3/s with a typical flow velocity of 13 m/s could be estimated, which lowered the water level in the ice-dammed lake by tens of meters (Smith and Fisher 1993; Norris et al. 2021). In this context, a climatic influence has also been speculated, as an inflow of 8.6 km^3/h of freshwater over a period of 2.5 months might have had an influence on ocean currents.

Corresponding considerations (Johnson and Lauritzen 1995) have been confirmed, especially with regard to ice dam outbursts with drainage into the north-western Atlantic, where melting sea ice might have supported the effect (Condron et al. 2020). During scientific exploratory drilling, rather coarse-grained components have repeatedly been discovered in the deep-sea deposits in the North Atlantic consisting mainly of clay (Ehlers 2022). The explanation of the origin of the so-called Heinrich Events posed difficulties, because gravelly deposits cannot be transported far out into the ocean from river mouths over long distances, as the low flow velocity does not provide sufficient drag force for further transport and deposition must occur not far from the mouth (cf. Chap. 2.3 for transport competence). However, transport on icebergs as ice-rafted debris (Fig. 5.19) offers a plausible explanation for the transport of coarse-grained sediments far out into the Atlantic.

Fig. 5.19 Transport of coarse-grained gravel on a drifting iceberg

Icebergs can break away from calving valley glaciers and ice sheets, carrying part of the moraine with them and eventually losing it when it melts far from the mainland. Corresponding deposits, which as the rock composition proves originate from North America, have been found off the coast of the British Isles, thus proving the range of this transport mechanism across the Atlantic Ocean. Now, on the other hand, it is not individual finds of single gravels but extensive sediment layers that prove that a whole armada of icebergs must have detached from the Laurentian Ice Sheet. Two considerations are pursued as explanations for the process responsible for this. Firstly, the large water surface of Lake Agassiz, which is open in summer—if the ice sheet had been closed, no shore lines could have been formed by waves on the shore—might have had an influence on the mass balance of the adjacent ice sheet in the form of increased precipitation from the water evaporating from the lake's surface (Hostetler et al. 2000). This increased precipitation will have led to an expansion of the ice sheet towards the Labrador Basin and the North Atlantic, where the ice is forced to float up, causing large amounts to calve and float as drifting icebergs on the ocean (Clarke et al.

Table 5.1 Selected outburst floods from Lake Agassiz into the Atlantic Ocean and Gulf of Mexico and related climate changes (Teller et al. 2002)

Climate change (cal. ^{14}C yr BP)	Volumen (km^3)	Lake level change (m)	Discharge (for duration of 1 year) (m^3/s)
Younger Dryas (12,900 yr)	9500	– 110	300,000
Pre-Boreal (11,700 yr)	9300	– 52	290,000
Pre-Boreal (11,200 yr)	5900	– 30	190,000
8200 Event (8400 yr)	163,000	– 770	5,200,000

1999). On the other hand, it can be argued that the outburst floods of Lake Agassiz, which passed under the ice sheet through Hudson Bay, must have affected the ice sheet, especially in the outlet area. A mechanical failure in the form of an open breach through the ice sheet hardly seems plausible due to its size. From the history of the lake it is possible to reconstruct when drainage events occurred and by what route they took. If we pay particular attention to those that flowed directly northeast into the North Atlantic (Table 5.1), it becomes clear that significant volumes of water were involved.

The water masses abruptly released into the North Atlantic had such a large volume that they influenced the ocean currents (cf. discussion in Naafs et al. 2013; Condron et al. 2020). This is especially true for the global thermohaline circulation. This global system of ocean currents is characterized as conveyor belt and is superordinate to the regional currents and driven by differences in the ocean water density due to temperature and salinity (Rahmstorf 2006). Specifically for the North Atlantic, this means that warm water from the Caribbean flows northwards at the surface, with Scandinavia's exceptionally mild climate for its latitude being caused by the Gulf Stream as part of this current. In the area of the European North Sea, the water cools noticeably, becoming heavier and sinking. In the process, it becomes part of the cool counter-current at depth and is carried south again (Fig. 5.20).

At the end of the last ice age, there were multiple climate setbacks in Central Europe during the rewarming (Table 5.1) (Ehlers 2022). These correlate remarkably well with striking outburst floods of Lake Agassiz into the North Atlantic. The enormous water masses released have dimensions comparable to ocean currents, where the smallest unit is a Sverdrup Sv (1 Sv = 1,000,000 m^3/s). If all these observations are combined, a coherent picture emerges, which illustrates that the outburst floods of Lake Agassiz at the end of the last ice age had far-reaching climatic consequences (Teller et al. 2002): with the abrupt release of the water masses from the lake, a lot of relatively light, since sweet water was sent into the North Atlantic. Due to its quantity, this water was able to interrupt the warm surface current into the European North Sea, thus causing a cold period in Europe. Witness to the range of the release of the water masses are the gravelly intercalations in the deep-sea deposits on the seabed of the North Atlantic. They prove that icebergs have drifted across the ocean on a large scale. In the process, the global climate system was disrupted in the form of a break in the thermohaline circulation in the North Atlantic, halting the trend to rewarming at the end of the Ice Age. In the end, global warming prevailed, so that the post-glacial period of the Holocene could also begin in Europe. A final setback occurred 8200 years ago, when the Laurentian Ice Sheet could no longer hold back the lake and the largest outburst flood ever abruptly emptied the lake completely (Clarke et al. 2004). This cold impulse also reached Europe (Von Grafenstein et al. 1998), but was too early to be included in the historical record.

5.3.2 Alluvial Fan Dynamics Rule Lake Levels—Lake Bonneville

Outside the glaciated areas, large parts of North America had a much wetter climate during the ice ages than today. This is particularly true of the semi-arid and nowadays desert areas of the south-west of the USA and its surroundings (Gillespie et al. 2004). Here, year-round water-bearing rivers flowed through a valley system that, due to

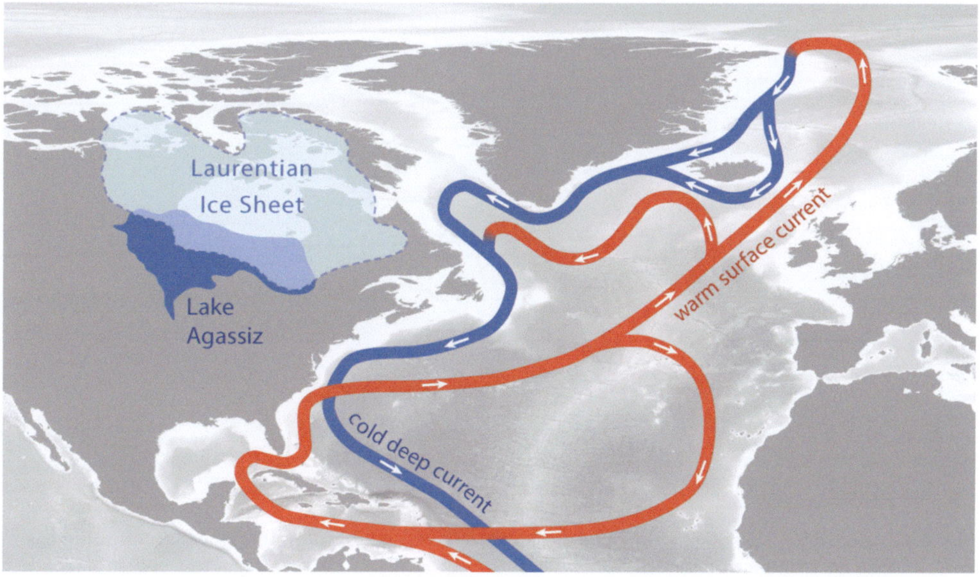

Fig. 5.20 Schematic of the present-day flows of the thermohaline circulation in the North Atlantic and location of Lake Agassiz dammed by the Laurentian Ice Sheet (after NASA Virtual Earth, Ruddiman 2001 and Teller et al. 2002)

Fig. 5.21 Shorelines of the glacial lake in the Great Salt Lake basin, Utah

its tectonic setting, did not offer drainage into the sea. As a result, water accumulated in the lowest basins, including Death Valley, whose southern inlet today only carries water in exceptionally wet winters. Water level elevation and timing of the maximum filling provides valuable palaeoclimate indications (Munroe and Laabs 2013) The basin of today's Great Salt Lake in the north of the US state of Utah also began to fill noticeably with water approximately 33,000 years ago (Janecke and Oaks 2011). 17,800 years ago, at an elevation of 1552 m above sea level, Lake Bonneville was formed with a depth of about 300 m as indicated by the elevation of the related shorelines (Fig. 5.21).

In the area of Red Rock Pass in southern Idaho, an alluvial fan of a stream flowing laterally from the mountains formed a dam. Due to the permanent water flow and rearrangement of the alluvial fan, the lake water was able to flow across the alluvial fan, but was not able to clear it out. This interplay of forces changed around 17,400 years ago when the water pressure of the lake was able to overcome the resistance of the alluvial fan and an outburst flood occurred (O'Connor et al. 2020). A volume of 4750 km^3

Fig. 5.22 Red Rock Pass in southern Idaho with the alluvial fan blocking the drainage from the basin and thereby forming Lake Bonneville (left, looking back into the Pleistocene lake basin)

emptied into the Snake River and flowed through its valley system into the Columbia River and finally reached the Pacific Ocean. The lake level was reduced by 103 m by corresponding incision in the area of the spillway (Fig. 5.22).

The outburst flood had a peak discharge of about 1,000,000 m^3/s in the area of the dam, still 935,000 m^3/s in the middle reaches of the Snake River and further reduced to about 600,000 m^3/s by retention losses at the border to Washington State (Jarrett and Malde 1987; O'Connor 1993). If the peak discharge is unrealistically simplified to be constant, the outburst flood lasted 8 weeks. In reality, the duration was certainly longer, with months probably being a realistic unit of time.

After the outburst flood and the reduction of the water level in the basin, a lake level stabilised at 1445 m above sea level, with a continuous outflow then ensured. This level is called Lake Provo and lasted until about 15,000 years before today. Due to decreasing precipitation, evaporation gained the upper hand again, as it still does today, the water balance became negative and the water level consequently dropped rapidly. Around 11,000 years before today, the current water level of the Great Salt Lake was reached at 1282 m, which has not changed much since then.

5.3.3 Textbook Example of Outburst Flood Studies—Lake Missoula

During the last ice age, named in North America Wisconsin Ice Age, valley glaciers formed a dense network in the Rocky Mountains and advanced into the foreland. One of these valley glaciers blocked the Clark Fork River in the north of the state of Idaho as it advanced, damming the lake named after the small Montana town Missoula in the valley system upstream (Fig. 5.23) (updated reviews by Baker et al. 2016; O'Connor et al. 2020; Bjornstad 2021; Ice Age Floods Institute 2024).

As the elevation of the shorelines (Fig. 5.24) indicates, the ice-dammed lake had a maximum water level at 1295 m a.s.l., resulting in a volume of 2184 km^3. In sheltered locations, lake sediments have been preserved, which show the characteristic seasonal layering of varves and also indicate interruption due to repeated drainage of the lake (Hanson et al. 2012).

Downstream the nearly 650 m high ice dam, outburst flood waves have repeatedly poured over the extensive basalt covers of the Columbia Plateau in eastern Washington State. The erosion of the loess cover on the basalts clearly traces the outburst floods (Fig. 5.25).

The Dry Falls at Coulee City, located west of Spokane, are marking the local transition from broad and rather shallow courses on the basalt surface to deeply incised gullies (Fig. 5.26). The gullies, called coulees, form an extensive network of about tens of more or less parallel courses, typically incised about 100 m deep. These gullies characterise the landscape of the Channelled Scabland, located between the coastal mountains in the west and the foothills of the Rocky Mountains in the east.

The Coulees converge again in southern Washington, where the Wallula Gap directs all runoff into the Columbia

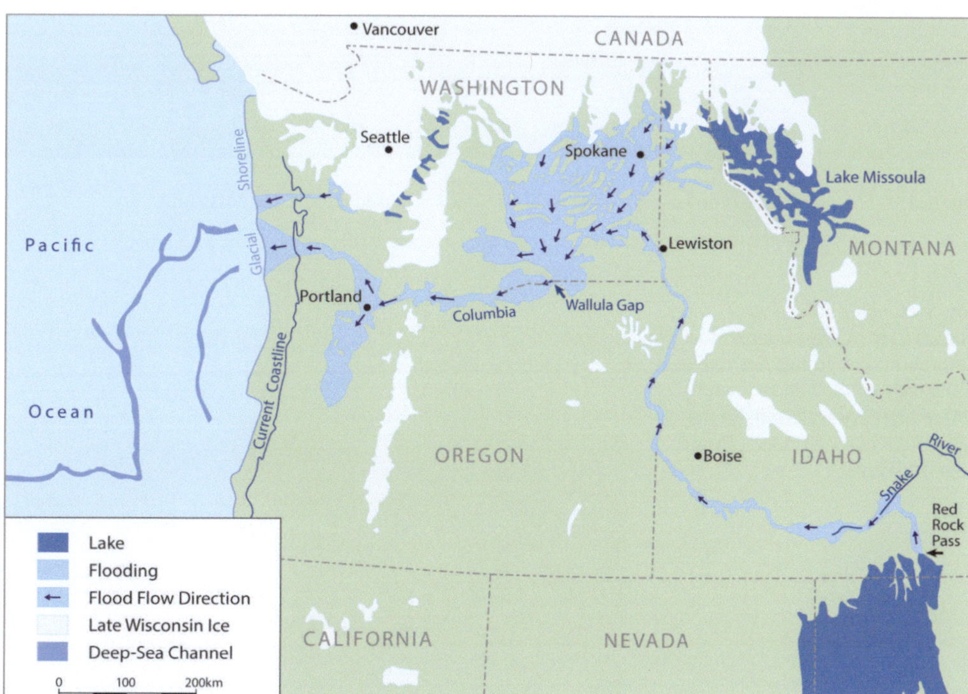

Fig. 5.23 Review of the ice-dammed Lake Missoula and the pathways of its outburst floods (modified after Herget 2003)

Fig. 5.24 Shorelines of the former ice-dammed lake on the slopes above the city of Missoula

Fig. 5.25 The eroded loess cover traces the outburst floods released from Lake Missoula

Fig. 5.26 Dry Falls near Coulee City in Washington State. Several incised gullies are visible at the edge of the approximately 100 m high dry waterfalls in the path of the outburst floods from Lake Missoula

River valley. Here, a backwater effect repeatedly occurred upstream the bottleneck, as evidenced by corresponding deposits and hydraulic calculations. Further runoff was then channelled through the Columbia valley into the Pacific Ocean. On the floor of the Pacific Ocean, the outburst floods have left sedimentary traces up to a distance of more than 1100 km from the mouth of the Columbia. On the one hand, sand intercalations are found in prominent landslides on the seafloor that clearly originate from the Columbia catchment (Zuffa et al. 2000), and in addition, the sudden input of large amounts of freshwater has also left traces in the species spectrum of microfossils (Lopes and Mix 2010). As in the Altai Mountains, fluvial gravel dunes, boulder blocks clearly transported by the Jökulhlaups of considerable size, and eroded scour structures whose dimensions exceed those described above are also found along the outburst flood trajectories as well as within the lake basin (Herget et al. 2013).

The last major outburst floods occurred between 17,500 and 14,500 years ago and minor refills without major outburst floods occurred even later (Hanson et al. 2012). No outburst floods occurred later than 13,700–13,400 year ago (O'Connor et al. 2020). There is also evidence that ice-dammed lakes formed in the region during previous ice ages and that outburst floods flowed through eastern Washington and down the Columbia valley before. Thus, deposits that are also related to outburst floods indicate an age before the last change in the Earth's magnetic field 780,000 years ago by their reversed magnetisation (Bjornstad et al. 2001). Considering the location of the region in the vicinity of the high mountains, which were undoubtedly repeatedly glaciated during the Pleistocene, the finding is not surprising and plausible. As a consequence, it should be expected that some of the coulees have a much older initial formation and have been enlarged by numerous jökulhlaups over time.

Estimating the magnitude of the outburst floods discharges is complicated by the topography with the separation into numerous coulees and the backwater in front of the Wallula Gap. The indication of flow velocities is therefore not meaningful. A physically based modelling of a drainage of the ice-dammed lake via englacial tubes in the ice resulted in values of 2,740,000–13,700,000 m^3/s within the area of the ice dam itself (Clarke et al. 1984). Assuming a complete failure of the ice dam, O'Connor and Baker (1992) determined a peak discharge of over 17,000,000 m^3/s for the region northeast of Spokane and of about 10,000,000 m^3/s in the Wallula Gap, advanced modelling confirmed this magnitudes (O'Connor et al. 2020). They conclude for a duration of the flood of a few days and do not exclude additional smaller events that may have occurred without complete opening of the ice dam by subglacial and englacial drainage. Benito (1997) confirms the estimation for the Wallula Gap based on his modelling in the Columbia Valley area. Advanced modelling approaches that also consider the process of dam failure, temporal changes in discharge respectively incision of the coulees provide comparable discharge magnitudes (Miyamoto et al.

Fig. 5.27 Layered stillwater deposits of Lake Missoula outburst floods near Wallula Gap. The layered deposits are called rhythmites and cover a dark ash layer from a Pleistocene eruption of Mount St. Helen volcano

2006; Alho et al. 2010; Denlinger and O'Connell 2010; Lehnigk and Larsen 2022).

As the extensive compilation of studies on the Lake Missoula outburst floods by the Ice Age Floods Institute (2024) documents, there is now a wealth of work on the subject, but some key aspects remain unresolved. Not really surprisingly, some dating results of outburst floods do not agree, especially when using different techniques and methods. The same applies to the uncertainties about the dam-break mechanism, because in order to generate the extreme peak discharges, it is difficult to imagine a discharge solely through englacial meltwater paths over the entire dam length of about 50 km. Furthermore, there is still an open question about the interpretation of the layered stillwater deposits (so-called rhythmites, e.g. Waitt 1985; Smith 1993), which can be found upstream of Wallula Gap (Fig. 5.27).

On the one hand, it is interpreted that each layer of the deposits represents a single jökulhlaup. Traces of bioturbation by animals burrowing in the soil serve as an argument, for which a longer period of time is needed between the deposition of the individual bands and the surface must dry out in the meantime. From the available dating for the period of the outburst floods, a periodic sequence of lake drains of 40–140 years results. Alternatively, a pulsating discharge is suggested. This is supported by the different lengths of the flood wave's paths through the various coulees as well as the plausibility of the temporary blockage of the outlet breach by collapsing ice of the glacier. The overall picture is further complicated when one considers that Lake Missoula was not the only source of flooding. Additional ice-dammed lakes existed west of Spokane and in the upper Columbia Valley as simplified illustrated in Fig. 5.23 (Kovanen and Slaymaker 2004). On the one hand, they may have generated by temporarily stored water from Lake Missoula, but on the other hand, they may have been a source of independent outburst floods (Atwater 1986; O'Connor et al. 2020). Furthermore, it must be considered that the Laurentian Ice Sheet was located to the north, which, with its continental extension, could only drain meltwater subglacially and that these meltwater courses had to follow the mountain relief in the area of the ice margin. It is therefore conceivable that large amounts of water were supplied to the Columbia River catchment from the ice, which may also have occurred as a high magnitude outburst flood out of subglacial reservoirs (Shaw et al. 2020). In addition to the water from Lake Missoula, glacial meltwater from the valley glaciers of the Rocky Mountains might have left traces in the region. Note, that at least once another flood drained through Columbia Valley is in any case certain due to the overflow of Lake Bonneville, which flowed into the Columbia via the Snake River valley (cf. previous chapter).

References

Abele G (1981) Trockene Massenbewegungen, Schlammströme und rasche Abflüsse - dominante morphologische Vorgänge in den chilenischen Anden. Mainzer Geographische Studien 23

Agatova AR, Nepop RK, Carling PA, Bohorquez P, Khazin LB, Zhdanova AN, Moska P (2020) Last ice-dammed lake in the Kuray basin, Russian Altai: New results from multidisciplinary research. Earth Sci Rev 205:103183

Alho P, Baker VR, Smith LN (2010) Paleohydraulic reconstruction of the largest Glacial Lake Missoula draining(s). Quatern Sci Rev 29:3067–3078

Arkhipov SA, Ehlers J, Johnson RG, Wright HE (1995) Glacial drainage towards the Mediterranean during the Middle and Late Pleistocene. Boreas 24:196–206

Astakhov VI (1991) The fluvial history of west Siberia. In: Starkel L, Gregory KJ, Thornes JB (eds) Temperate palaeohydrology. Wiley, Chichester, pp 381–392

ASTER GDEM (2009) NASA/METI/AIST/Japan Space systems, and U.S./Japan ASTER Science Team. ASTER Global Digital Elevation Model. 2009, Distributed by NASA EOSDIS Land Processes DAAC. https://doi.org/10.5067/ASTER/ASTGTM.002

Atwater B (1986) Status of glacial Lake Columbia during the last floods from glacial Lake Missoula. Quatern Res 27:182–201

Baker VR (1988) Cataclysmic processes in geomorphological systems. Zeitschrift Für Geomorphologie, Supplement-Band 67:25–32

Baker VR (2009) The Channeled Scabland—a retrospective. Annu Rev Earth Planet Sci 37:393–411

Baker VR, Carling PA (2020) Global late Quaternay megafloods. https://doi.org/10.1016/B978-0-12-409548-9.12459-5

Baker VR, Benito G, Rudoy AN (1993) Paleohydrology of Late Pleistocene superflooding, Altay Mountains, Siberia. Science 259:348–350

Baker VR, Bjornstad BN, Gaylord DR, Smith GA, Meyer SE, Alho P, Breckenridge RM, Sweeney MR, Zreda M (2016) Pleistocene megaflood landscapes of the Channeled Scabland. In: Lewis RS, Schmidt KL (eds) Exploring the geology of the inland northwest. Geological Society of America Field Guide, vol 41, pp 1–73

Batbaartar J, Gillespie AR (2016a) Outburst floods of the Maly Yenisei—Part I. Int Geol Rev 58:1723–1752

Batbaartar J, Gillespie AR (2016b) Outburst floods of the Maly Yenisei—Part II—new age constraints from Darhad basin. Int Geol Rev 58:1753–1779

Batbaartar J, Gillespie AR, Fink D, Matmon A, Fujioka T (2018) Asynchronous glaciations in arid continental climate. Quatern Sci Rev 182:1–19

Benito G (1997) Energy expenditure and geomorphic work of the cataclysmic Missoula flooding in the Columbia River gorge, USA. Earth Surf Proc Land 22:457–472

Bjornstad B (2008) Technical bibliography—scientific papers and books about the ice age floods. Ice Age Flood Institute: http://www.iafi.org/techpubsalpha.html. 16 Apr 2012

Bjornstad BN (2021) Ice age floodscapes of the Pacific Northwest—a photographic exploration. Springer, Cham

Bjornstad BN, Fecht KR, Pluhar CJ (2001) Long history of pre-Wisconsin, ice age cataclysmic floods—evidence from southeastern Washington State. J Geol 109:695–713

Bohorquez P, Carling PA, Herget J (2015) Dynamic simulation of the catastrophic drainage within the basin of late-Pleistocene glacial-lake Kuray-Chuja, Altai Mountains, Southern Siberia. International Geological Review. https://doi.org/10.1080/00206814.2015.1046956

Bohorquez P, Jimenez-Ruiz PJ, Carling PA (2019) Revisiting the dynamics of catastrophic late Pleistocene glacial-lake drainage, Altai Mountains, central Asia. Earth Sci Rev 197:102892

Brunnberg L (1995) The Baltic Ice Lake. Quatern Int 28:177–178

Busschers F, van Balen RT, Cohen KM, Kasse C, Weerts HJT, Walinga J, Bunnik FPM (2008) Response of the Rhine-Meuse fluvial system to Saalian ice-sheet dynamics. Boreas 37:377–398

Butvilovsky VV (1993) The last glaciation and the holocene palaeogeography of the Altai Mountains—a cataclysmic model. Tomsk (in Russian)

Carling PA (1996) Morphology, sedimentology and palaeohydraulic significance of large gravel dunes, Altai Mountains, Siberia. Sedimentology 43:647–664

Carling PA, Fan X (2020) Particle comminution defines megaflood and superflood energetics. Earth Sci Rev 2024:103087

Carling PA, Kirkbride AD, Parnachov SV, Borodavko PS, Berger GW (2002) Late-Quaternary catastrophic flooding in the Altai Mountains of south-central Siberia: a synoptic overview and an introduction to flood deposits sedimentology. In: Martini PI, Baker VR, Garzon G (eds) Flood and megaflood processes and deposits: recent and ancient examples. International Association of Sedimentologists, Special Publication, vol 32, pp 17–35

Carling PA, Villanueva I, Herget J, Wright N, Borodavko P, Morvan H (2010) Unsteady 1-D and 2-D hydraulic models with ice-dam break for Quaternary megaflood, Altai Mountains, southern Siberia. Global Planet Change 70:24–34

Carling PA, Bohorquez P, Fan X (2020) Hydraulic control on the development of megaflood runup deposits. Geomorphology 361:107203

Clarke GKC, Mathews WH, Pack RT (1984) Outburst floods from glacial Lake Missoula. Quatern Res 22:289–299

Clarke GKC, Marshall SJ, Hillaire-Marcel C, Bilodeau G, Veiga-Pires C (1999) A glaciological perspective on Heinrich Events. In: Clark PU, Webb RS, Keigwin LD (eds) Mechanisms of global climate change at millennial time scales. American Geophysical Union, Washington, pp 243–262

Clarke GKC, Leverington DW, Teller JT, Dyke AS (2004) Paleohydraulics of the last outburst flood from glacial Lake Agassiz and the 8200 BP cold event. Quatern Sci Rev 23:389–407

Cohen KM, Gibbard PL (2019) Global chronostratigraphical correlation table for the last 2.7 million years. Quatern Int 500:20–31

Cohen KM, Gibbard PL (2022) Global chronostratigraphical correlation table for the last 2.7 million years v.2022 (Poster version). Mendeley Data, vol 5. https://doi.org/10.17632/dtsn3xn3n6.5

Condron A, Joyce AJ, Bradley RS (2020) Arctic sea ice export as a driver of deglacial climate. Geology 48:395–399

Davis R, 13 co-authors (2023) Middle Palaeolithic occupation of the southern North Sea Basin—evidence from the sandscaping sediments emplaced on the beach between Bacton and Walcott, Norfolk, UK. J Quatern Sci. https://doi.org/10.1002/jqs.3524

Denlinger RP, O'Connell DRH (2010) Simulations of cataclysmic outburst floods from Pleistocene glacial Lake Missoula. Geol Soc Am Bull 122:678–689

Ehlers J, Gibbard PL (eds) (2004) Quaternary glaciations—extent and chronology, part I–III. Elsevier, Amsterdam

Ehlers J (2022) The ice age. Springer, Berlin

Ehlers J, Gibbard PL, Hughes PD (eds) (2011) Quaternary glaciations—extent and chronology, a closer look. Elsevier, Amsterdam

Fisher TG (2004) River Warren boulders, Minnesota, USA—catastrophic paleoflow indicators in the southern spillway of glacial Lake Agassiz. Boreas 33:349–358

García-Moreno D, Gupta S, Collier JS, Oggioni F, Vanneste K, Trentesaux A, Verbeeck K, Versteeg W, Jomard H, Camelbeeck T, De Batist M (2019) Middle—Late Pleistocene landscape evolution of the Dover Strait inferred from buried and submerged erosional landforms. Quatern Sci Rev 203:209–232

Gillespie AR, Porter SC, Atwater BF (eds) (2004) The Quaternary period in the United States. Elsevier, Amsterdam

Grosswald MG (1999) Cataclysmic megafloods in Eurasia and the polar ice sheets. Scientific World, Moskau (in Russisch)

Grosswald MG, Rudoy AN (1996) Quaternary glacier-dammed lakes in the mountains of Siberia. Polar Geogr 20:180–198

Grosswald MG, Hughes TJ (2002) The Russian component of an Arctic ice sheet during the Last Glacial Maximum. Quatern Sci Rev 21:121–146

Gupta S, Collier JS, Palmer-Felgate A, Potter G (2007) Catastrophic flooding origin of shelf valley systems in the English Channel. Nature 448:342–345

Hanson MA, Lian OB, Clague JJ (2012) The sequence and timing of large late Pleistocene floods from glacial Lake Missoula. Quatern Sci Rev 31:67–81

Harrison SDES, Glasser NF (2018) Late Quaternary meltwater pulses and sea level change. J Quatern Sci https://doi.org/10.1002/jqs.3070

Herget J (1998) Temporäre Entwässerungsbahnen am Südrand der Westfälischen Tieflandsbucht - ein Szenario. Bochumer Geographische Arbeiten Sonderreihe 13:23–30

Herget J (2002) Sedimentary traces as indicator of temporary ice-marginal channels in the Westphalian Bight, Germany. In: Martini PI, Baker VR, Garzón G (eds) Flood and megaflood processes and deposits: recent and ancient examples. Special Publications of the International Association of Sedimentologists, vol 32, pp 283–290

Herget J (2003) Eisstausee-Ausbrüche - Ursache für katastrophale Hochwasser. Geogr Rundsch 55–2:14–20

Herget J (2005) Reconstruction of Pleistocene ice-dammed lake outburst floods in Altai-Mountains. Geological Society of America, Special Publication, Siberia, p 386

Herget J (2010) Extreme Hochwasser in der jüngeren Erdgeschichte. Geogr Rundsch 3(2010):26–33

Herget J, Euler T, Roggenkamp T, Zemke J (2013) Obstacle marks as palaeohydraulic indicators of Pleistocene megafloods. Hydrol Res 44:300–317

Herget J, Agatova AR, Carling PA, Nepop RK (2020) Altai megafloods—the temporal context. Earth Sci Rev 200:102995

Hijma MP, Cohen KM, Roebroeks W, Westerhoff WE, Busschers FS (2012) Pleistocene Rhine-Thames landscapes—geological background for hominin occupation of the southern North Sea region. J Quatern Sci 27:17–39

Hostetler SW, Bartlein PJ, Clark PU, Small EE, Solomon AM (2000) Simulated influences of Lake Agassiz on the climate of central North America 11,000 years ago. Nature 405:334–337

Hughes ALC, Gyllencreutz R, Lohne ØS, Mangerud J, Svendsen JI (2016) The last Eurasian ice sheets—a chronological database and time-slice reconstruction DATED-1. Boreas 45:1–45

Hyttinen O, Kotilainen A, Salonen V-P (2011) Acoustic evidence of a Baltic Ice Lake drainage debrite in the northern Baltic Sea. Mar Geol 284:139–148

Ice Age Floods Institute (ed) (2024) Ice Age Floods Institute—4 states, 11 chapters. https://www.iafi.org. 30 May 2024

Jakobsson M, Björck S, Alm G, Andrén T, Lindeberg G, Svensson N-O (2007) Reconstructing the Younger Dryas ice dammed lake in the Baltic Basin—bathymetry, area and volume. Global Planet Change 57:355–370

Janecke SU, Oaks RQ (2011) Reinterpreted history of latest Pleistocene Lake Bonneville. In: Lee J, Evans JP (eds) Geologic field trips to the basin and range, Rocky Mountains, Snake River Plain, and Terranes of the US Cordillera. Geological Society of America Field Guide, vol 21, pp 195–222

Jarrett RD, Malde HE (1987) Paleodischarge of the late Pleistocene Bonneville Flood, Snake River, Idaho, computed from new evidence. Geol Soc Am Bull 99:127–134

Johnson RG, Lauritzen S-E (1995) Hudson Bay—Hudson Strait jökulhlaups and Heinrich events—a hypothesis. Palaeogeogr Palaeoclimatol Palaeoecol 117:123–137

Johnson MD, Öhrling C, Bergström A, Dreyer Isaksson O, Pizarro Rajala E (2022) Geomorphology and sedimentology of features formed at the outlet during the final drainage of the Baltic Ice Lake. Boreas 51:20–40

Junge FW (1998) Die Bändertone Mitteldeutschlands und angrenzender Gebiete. Altenburger Naturwissenschaftliche Forschungen 9

Kehew AE, Lord ML, Kozlowski AL, Fisher TG (2009) Proglacial megaflooding along the margins of the Laurentide Ice Sheet. In: Burr DM, Carling PA, Baker VR (eds) Megaflooding on earth and mars. Cambridge University Press, Cambridge, pp 104–127

Knies J, Vogt C (2003) Freshwater pulses in the eastern Arctic Ocean during Saalian and Early Weichselian ice-sheet collapse. Quatern Res 60:243–251

Komatsu G, Arzhannikov SG, Gillespie AR, Burke RM, Miyamoto H, Baker VR (2009) Quaternary paleolake formation and cataclysmic flooding alaong the upper Yenisei River. Geomorphology 104:143–164

Kovanen DJ, Slaymaker O (2004) Glacial imprints of the Okanogan Lobe, southern margin of the Cordilleran Ice Sheet. J Quat Sci 19:547–565

Krivonogov S, Zolnikov I, Novikov I, Deev E (2017) Giant glaciogenic floods in the Altai—geomorphological, geological and hydrological aspects—guidebook for field excursion at the 14th international workshop on present earth surface processes and longterm environmental changes in East Eurasia, 15–21 Sept 2017. Novosibirsk

Lampe R (2002) Post-glacial water-level variability along the south Baltic coast—a short review. Greifswalder Geographische Arbeiten 27:13–19

Lehnigk KE, Larsen IJ (2022) Pleistocene megaflood discharge in Grand Coulee, Channeled Scabland, USA. J Geophys Res Earth Surf 127:e2021JF006135

Lewis CFM, Teller JT (2007) North American late-quaternary meltwater and floods to the oceans—evidence and impact. Palaeogeogr Palaeoclimatol Palaeoecol 246:1–161

Liedtke H (1981) Die Nordischen Vereisungen in Mitteleuropa. Forschungen zur Deutschen Landeskunde 204

Lopes C, Mix AC (2010) Pleistocene megafloods in the northeast Pacific. Geology 37:79–82

Mangerud J, Jakobsson M, Alexanderson H, Astakhov V, Clarke GKC, Henriksen M, Hjort C, Krinner G, Lunkka J-P, Möller P, Murray A, Nikolskaya O, Saarnisto M, Svendsen JI (2004) Ice-dammed lakes and rerouting of the drainage of northern Eurasia during the last glaciation. Quatern Sci Rev 23:1313–1332

Mangerud J, Astakhov V, Jakobsson M, Svendsen JI (2001) Huge iceage lakes in Russia. J Quat Sci 16:773–777

Margold M, Jansen JD, Gurinov AL, Codilean AT, Fink D, Preusser F, Reznichenko NV, Mifsud C (2016) Extensive glaciation in Transbaikalia, Siberia, at the Last Glacial Maximum. Quatern Sci Rev 132:161–174

Margold M, Jansen J, Codilean A, Preusser F, Gurinov A, Fujioka T, Fink D (2018) Repeated megafloods from glacial Lake Vitim, Siberia, to the Arctic Ocean over the past 60,000 years. Quatern Sci Rev 187:41–61

Meinsen J, Winsemann J, Weitkamp A, Landmeyer N, Lenz A, Dölling M (2011) Middle Pleistocene (Saalian) lake outburst floods in the Münsterland Embayment (NW Germany)—impacts and magnitudes. Quatern Sci Rev 30:2597–2625

Miyamoto H, Itoh K, Komatsu G, Baker VR, Dohm JM, Tosaka H, Sasaki S (2006) Numerical simulations of large-scale cataclysmic floodwater—a simple depth-averaged model and an illustrative application. Geomorphology 76:179–192

Munroe JS, Laabs BJC (2013) Temporal correspondence between pluvial lake highstands in the southwestern US and Heinrich Event 1. J Quat Sci 28:49–58

Nørgaard J, Margold M, Jansen JD, Kurbanov R, Szuman I, Lund Andersen J, Olsen J, Faurschou Knudsen M (2023) Absence of large-scale ice masses in central Northeast Siberia during the Late Pleistocene. Geophysical Research Letters 50:e2023GL103594

Norris SL, Garcia-Castellanos D, Jansen JD, Carling PA, Margold M, Woywitka RJ, Froese DG (2021) Catastrophic drainage from the northwestern outlet of glacial Lake Agassiz during the Younger Dryas. Geophys Res Lett 48:e2021GL093919

O'Connor JE (1993) Hydrology, hydraulics, and geomorphology of the Bonneville Flood. Geological Society of America Special Paper 274

O'Connor JE, Baker VR (1992) Magnitudes and implications of peak discharges from glacial Lake Missoula. Geol Soc Am Bull 104:267–279

O'Connor JE, Baker VR, Waitt RB, Smith LN, Cannon CM, George DL, Denlinger RP (2020) The Missoula and Bonneville floods—a review of ice-age megafloods in the Columbia River basin. Earth Sci Rev 208:103181

Rahmstorf S (2006) Thermohaline ocean circulation. In: Elias SA (ed) Encyclopedia of quaternary sciences. Elsevier, Amsterdam, pp 739–750

Reuther A, Herget J, Ivy-Ochs S, Borodavko P, Kubik PW, Heine K (2006) Constraining the timing of the most recent cataclysmic flood event from ice-dammed lakes in the Russian Altay Mountains, Siberia, using cosmogenic in-situ ^{10}Be. Geology 34:913–916

Ruddiman WF (2001) Earth's climate—past and future. Freeman, New York

Rudoy AN (2002) Glacier-dammed lakes and geological work of glacial superfloods in the Late Pleistocene, southern Siberia, Altai Mountains. Quatern Int 87:119–140

Shaw J, Gilbert RG, Sharpe DR, Lesemann J-E, Young RR (2020) The origins of s-forms—form similarity, process analogy, and links to high-energy, subglacial meltwater flows. Earth Sci Rev 200:102994

Sidorchuk AY, Panin AV, Borisova OK (2011) Surface runoff to the Black Sea from the East European Plain during the Last Glacial Maximum—late glacial time. Geol Soc Am Spec Pap 473:1–25

Skupin, K. & J.G. Zandstra (2011): Gletscher der Saale-Kaltzeit am Niederrhein. Krefeld (Geologischer Dienst NRW).

Smith AJ (1985) A catastrophic origin for the palaeovalley system of the eastern English Channel. Mar Geol 64:65–75

Smith GA (1993) Missoula flood dynamics and magnitudes inferred from sedimentology of slack-water deposits on the Columbia Plateau, Washington. Geol Soc Am Bull 105:77–100

Smith DG, Fisher TG (1993) Glacial Lake Agassiz—the northwestern outlet and paleoflood. Geology 21:9–12

Teller JT (2001) Formation of large beaches in an area of rapid differential isostatic rebound—the three-outlet control of Lake Agassiz. Quatern Sci Rev 20:1649–1659

Teller JT, Leverington DW, Mann JD (2002) Freshwater outbursts to the oceans from glacial Lake Agassiz and their role in climate change during the last deglaciation. Quatern Sci Rev 21:879–887

Thome KN (1983) Gletschererosion und -akkumulation im Münsterland und angrenzenden Gebieten. Neues Jb Geol Paläontol Abh 166:116–138

Tizzard L, Bicket AR, Benjamin J, De Loecker D (2014) A Middle Palaeolithic site in the southern North Sea—investigating the archaeology and palaeogeography of Area 240. J Quatern Sci 29:698–710

Toucanne S, Zaragose S, Bourillet JF, Cremer M, Eynaud F, van Vliet-Lanoë B, Penaud A, Fontanier C, Turon JL, Cortijo E, Gibbard PL (2009) Time of massive Fleuve Manche discharges over the last 350kyr—insights into the European ice-sheet oscillations and the European drainage network from MIS 10 to 2. Quatern Sci Rev 28:1238–1256

Toucanne S, Zaragose S, Bourillet J-F, Marieu V, Cremer M, Kageyama M, van Vliet-Lanoë B, Eynaud F, Turon J-L, Gibbard PL (2010) The first estimation of Fleuve Manche palaeoriver discharge during the last deglaciation—evidence for Fennoscandian ice sheet meltwater flow in the English Channel ca 20–18 ka ago. Earth Planet Sci Lett 290:459–473

Tripsanas EK, Bryant WR, Slowey NC, Bouma AH, Karageorgis AP, Berti D (2007) Sedimentological history of Bryant Canyon area, northwest Gulf of Mexico, during the last 135 kyr—a proxy record of Mississippi River discharge. Palaeogeogr Palaeoclimatol Palaeoecol 246:137–161

Vink A, Steffen H, Reinhardt L, Kaufmann G (2007) Holocene relative sea-level change, isostatic subsidence and the radial viscosity structure of the mantle of northwest Europe (Belgium, the Netherlands, Germany, southern North Sea). Quatern Sci Rev 26:3249–3275

Von Grafenstein UH, Erlenkeuser J, Müller JJ, Johnsen S (1998) The cold event 8200 year ago documented in oxygen isotope records of precipitation in Europe and Greenland. Clim Dyn 14:73–81

Waitt RB (1985) Case for periodic, colossal jökulhlaups from Pleistocene glacial Lake Missoula. Geol Soc Am Bull 96:1271–1286

Weckwerth P, Wysota W, Piotrowski JA, Adamczyk A, Krawiec A, Dąbrowski M (2019) Late Weichselian glacier outburst floods in North-Eastern Poland—landform evidence and palaeohydraulic significance. Earth Sci Rev 194:216–233

Winsemann J, Brandes CH, Polom U, Weber CH (2011) Depositional architecture and palaeogeographic significance of Middle Pleistocene glaciolacustrine ice marginal deposits in northwestern Germany—a synoptic review. E&G Quatern Sci J 60:212–235

Winsemann J, Alho P, Laamanen L, Goseberg N, Lang J, Klostermann J (2015) Flow dynamics, sedimentation and erosion of glacial lake outburst floods along the Middle Pleistocene Scandinavian Ice Sheet (northern central Europe). Boreas. https://doi.org/10.1111/bor.12146

Zolnikov ID, Deev EV, Nazarov DV, Kotler SA (2015) Comparative analysis of megaflood deposits and alluvium of the Chuya and Katun river valleys (Gorny Altai). Russ Geol Geophys 56:1162–1172

Zuffa GG, Normark WR, Serra F, Brunner CA (2000) Turbidite megabeds in an oceanic rift valley recording jökulhlaups of Late Pleistocene glacial lakes of the western United States. J Geol 108:253–274

Further Reading

Blocked river courses by Pleistocene ice-shields both in general and specifically in Central Europe and presented in review by:

Panin A, Astakhov VI, Lotisari E, Komatsu G, Lang J, Winsemann J (2020) Middle and Late Quaternary glacial lake-outburst floods, drainage diversions and reorganization of fluvial systems in northwestern Eurasia. Earth Sci Rev 201:103069

Winsemann J, Lang J (2020) Flooding Northern Germany—impacts and magnitudes of Middle Pleistocene glacial-lake outburst floods. In: Herget J, Fontana A (eds) Palaeohydrology—traces, tracks and trails of extreme events. Springer, Cham, pp 29–47

A general review on the Quaternary development of the Baltic Sea is given by:

Björk S (1995) A review of the history of the Baltic Sea, 13.8-8.0 ka BP. Quatern Int 27:19–40

Environmental consequences of the separation of the British Islands from the European Mainland during the Pleistocene are illustrated by:

Gibbard PL, Lautridou JP (2003) The Quaternary history of the English Channel. J Quat Sci 18:195–371

Preece RC (ed) (1995) Island Britain—a Quaternary perspective. Geological Society of London Special Publication 96

A review of the research history of Pleistocene megafloods in northern Eurasia including studies previously published in Russian language only is provided by:

Komatsu G, Baker VR, Arzhannikov SG, Gallagher R, Arzhannikova AV, Murana A, Oguchi T (2015) Catastrophic flooding, palaeolakes, and late Quaternary drainage reorganization in northern Eurasia. Int Geol Rev 58:1693–1722

In addition to several publications on specific details respectively environmental and flood conditions still under controversial discussion, some review papers provide access to further details including additional studies not referenced here:

Bohorquez P, Jimenez-Ruiz PJ, Carling PA (2019) Revisiting the dynamics of catastrophic late Pleistocene glacial-lake drainage, Altai Mountains, central Asia. Earth Sci Rev 197:102892

Carling PA (2013) Freshwater megaflood sedimentation: What can we learn about generic processes? Earth-Sci Rev 125:87–113

Carling PA, Martini P, Herget J, Borodavko P, Parnachov S (2009) Megaflood sedimentary valley fill—Altai Mountains, Siberia. In: Burr DM, Carling PA, Baker VR (eds) Megaflooding on earth and mars. Cambridge University Press, Cambridge, pp 243–264

Carling PA, Knaapen K, Borodavko P, Herget J, Koptev I, Huggenberger P, Parnachov S (2011) Palaeoshorelines of glacial Lake Kuray–Chuya, south-central Siberia—form, sediments and process. In: Martini IP, French HM, Pérez Alberti A (eds) Ice-marginal and periglacial processes and sediments. Geological Society London Special Publication 354, pp 111–128

Heinrich Events, ice-rafted debris and the influence of drainage from North America on ocean currents and global climate are explained and discussed in further detail in:

Condron A, Joyce AJ, Bradley RS (2020) Arctic sea ice export as a driver of deglacial climate. Geology 48:395–399

Naafs BDA, Hefter J, Stein R (2013) Millennial-scale ice rafting events and Hudson Strait Heinrich(-like) Events during the late Pliocene and Pleistocene—a review. Quatern Sci Rev 80:1–28

Teller JT, Leverington DW, Mann JD (2002) Freshwater outbursts to the oceans from glacial Lake Agassiz and their role in climate change during the last deglaciation. Quatern Sci Rev 21:879–887

A review on development of the spillways, field evidence and lake levels is presented with comments on remaining open questions by

Fisher T (2020) Megaflooding associated with glacial Lake Agassiz. Earth Sci Rev 201:102974

Additional details and a chronology of the levels of Lake Bonneville and successors in further details is presented by

Oviatt CG (2015) Chronology of Lake Bonneville, 30,000 to 10,000 yr B.P. Quatern Sci Rev 110:166–171

A substantial compilation of papers, maps, and materials like photos and videos on the ice-dammed lake outburst floods in north-western USA is provided by

Ice Age Floods Institute (ed) (2024) Ice Age Floods Institute—4 states, 11 chapters. https://www.iafi.org. 30 May 2024

Inspiring comments and hints on open questions in the context of both Lake Bonneville and Lake Missoula floods are compiled in

O'Connor JE, Baker VR, Waitt RB, Smith LN, Cannon CM, George DL, Denlinger RP (2020) The Missoula and Bonneville floods—a review of ice-age megafloods in the Columbia River basin. Earth Sci Rev 208:103181

6 Refill of Oceanic Basins in Earth's History

The refill of oceanic basins in the Earth's recent history since the end of the Tertiary represent floods of a special magnitude. Such floods do not manifest themselves through a temporarily increased discharge, but as a rule represent a distinct event. Compared to the lake outbursts described before, the direction of flow is reversed due to the inflow into the basin instead of draining it. At the same time, the duration of the floods also changes, which is typically measured in years according to the size of the basin. Questions also arise not only about the cause of the flood when it fills, but also about how, when and why a previous emptying occurred in the first place.

What sounds very theoretical and abstract at first, becomes more vivid when looking at the following examples, because on the one hand, we are dealing with the most famous and well-known flood of all: the great deluge. On the other hand, the largest flood that can ever be proven on earth has to be mentioned in this context.

6.1 Black Sea

Today, the Black Sea is an almost closed sub-basin of the Mediterranean Sea, which is connected to the main basin via the Aegean Sea, the straits of the Dardanelles, the Sea of Marmara and ultimately the Bosporus (Figs. 6.1 and 6.2). With a minimum depth of 30 m at the Bosporus and 70 m at the Dardanelles, it became an inland sea during the last ice age, when sea-level subsidence was also as low as −120 m in the Mediterranean (Shackleton 1987). Since the regular water supply via the main tributary Rivers Danube, Dnjpro, Dnjester and Don could not compensate for the evaporation losses also during glacial times, the water level dropped to as low as −123 m (Ryan et al. 2003). However, the maximum depth of 2212 m resulting from tectonic subsidence ensured that a residual sea was preserved, even if the sea surface area was much smaller because of the wide shallow shelf (Figs. 6.1 and 6.2).

While a low water level in the Black Sea generally does not give rise to discussion, the depth of the water level and especially the replenishment of the basin is subject of intensive research. Four contrasting working hypotheses have been developed, which can be differentiated as follows:

(1) Catastrophic water intrusion from the Mediterranean,
(2) Filling by inflow from the Caspian Sea,
(3) Continuous refill with bidirectional exchange with the Mediterranean,
(4) Continuous refill with permanent flow into the Mediterranean.

These four explanatory approaches presented below in subchapters postulate a different origin of the water for the replenishment of the Black Sea Basin and indicate different points in time for this. Each of the hypotheses can also come up with valid findings that are not easy to refute. There is not enough space here to evaluate and discuss the above arguments of the various working hypotheses in detail, even though the catastrophe sceptics seem to predominate (numerically) in the meantime. A compilation of the different findings, conclusions and argumentations is offered in the comprehensive book edited by Yanko-Hombach et al. (2007a) and in an updated version in Yanko-Hombach and Gilbert (2018). The contrary concepts and related evidence are presented by e.g. Yanko-Hombach and Yilmaz (2007), Yanko-Hombach and Smyntyna (2009) and Buynevich et al. (2011) on the one hand side and e.g. Ryan et al. (1997), Ryan and Pitman (1998) and Yanchilla et al. (2017) on the other. A generalised and schematised overview is given below.

6.1.1 Catastrophic Water Intrusion into the Black Sea

During the last glacial maximum at about 20,000 years ago, the Black Sea had a temporary high water level due to additional water from northeast European ice-dammed lakes via the Volga River, the Caspian Sea and the Manych

J. Herget, *Flood Reconstruction*, Springer Natural Hazards, https://doi.org/10.1007/978-3-031-66815-9_6

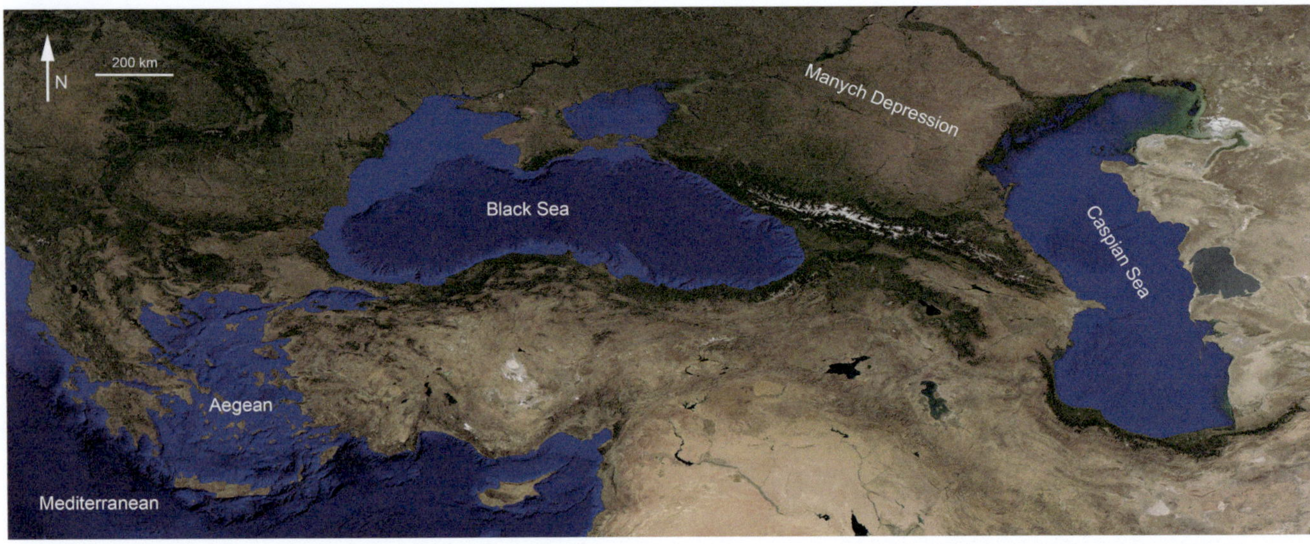

Fig. 6.1 Bathymetric overview of the Black Sea and adjacent areas (satellite image from NASA Visual Earth)

Fig. 6.2 Bathymetric and topographic overview of the transition area between the Mediterranean and the Black Sea (satellite image from NASA Visual Earth)

Depression (Fig. 5.5). At times (16,000–14,700 and 11,000–10,000 years before present) water flowed from the Black Sea into the lower lying Mediterranean Sea (Ryan et al. 2003). Subsequently, the water level dropped to lows of up to −123 m in synchronicity with the re-warming and the diminishing of this additional water source. Evidence for lows in the Black Sea at the end of the last glaciation is provided by plant roots, dry cracks and algal remains, as well as incised river channels in the estuaries of the Black Sea. While the water level in the Mediterranean is already rising again, that of the Black Sea lags behind, but a water overflow is prevented by a sediment dam in the area of the Bosporus. Around 8400 years before today, this dam finally breaks. At this time, the water level in the Mediterranean Sea was −15 m compared to today, but that of the Black Sea was still −150 m. Initial estimates assume inflows of > 50 km^3 per day (equivalent to > 500,000 m^3/s) and rates of rise of the water level in the order of decimetres per day (Ryan et al 1997). Subsequent modelling concretised this first estimate to inflows of a maximum of 175,000 m^3/s, taking into account the maximum flow capacity of the strait, resulting in a refilling time of at least 8 years (Myers et al. 2003), or 58,000 m^3/s, resulting in 34 years until the water levels are equalised (Sidall et al. 2004). Echosounder measurements document structures at the bottom of the basin, which are attributed to strong currents during the basin filling. Another important argument is an abrupt change in the strontium isotope spectrum, which indicates an abrupt inflow of water from the Mediterranean Sea.

These investigations have attracted particular attention because the authors have drawn more far-reaching conclusions for the settlements in the region. The plausible-sounding hypothesis is that with the very rapid refilling of the sea basin, people who soon settled on the shore of the sea were also affected by the rise of the sea. A small rise in sea level leads to a wide expanse on the shallow coast. Escape and protection possibilities do not exist in the shallow area, because it is the dry older seabed of the Black Sea, which was previously levelled by the wave action of the water, and the former mainland areas are far away, especially in the northwest (Fig. 6.1). This experience must have been traumatising and is seen as the starting point for the biblical deluge myth (Ryan and Pitman 1998) (cf. below for details).

6.1.2 Refill Through the Manych Depression from the Caspian Sea

In this explanatory approach, a late glacial replenishment of the Black Sea by inflow of glacial meltwater from Central Asia, namely the Aral Sea, is assumed (Chepalyga 2007). In addition, when the ice sheet melted in the north, the rivers carried large amounts of additional meltwater southwards through eastern Europe, as indicated by the preserved large meanders in the river catchments (cf. Sect. 2.4). In the form of a cascade of basins and spillways, a connection is seen from the Aral Sea via the Caspian Sea into the Black Sea and further on via a precursor of today's Sea of Marmara into the Mediterranean Sea. This cascade is supported by dating as in active use in the period 17,000–13,000 years before present. Specific clay deposits and endemic mollusc species are seen as evidence of the connections. Interbasin discharges are given as 40,000–60,000 m^3/s, only the overflow from the Aral Sea had a discharge of 3000 m^3/s. For the Black Sea, an increase of 60–70 m is given due to repeated overflows from the Caspian Sea in the period between 16,000–14,000 years before today, which corresponds to an increase of about 3 cm per year. The remaining residual filling of the basin occurred later by the rise of the Mediterranean Sea level, but then no longer had catastrophic dimensions.

The parallels to the drainage of the Northeast European and West Siberian ice-dammed lakes during the last ice age are obvious (cf. Sect. 5.2). The focus of the concept is on the dynamics and development of the water level in the Caspian Sea, while the development in the Black Sea is considered more as a consequence of this. For the Caspian Sea, the development is seen as catastrophic and also impulsive for the Flood myth. Here, too, it is less the dramatic rate of water level rise than the resulting speed of sea level spread on the populated shelf area.

6.1.3 Continuous Refill with Bidirectional Exchange with the Mediterranean

Testing the above hypotheses, Yanko-Hombach et al. (2007b) conclude that the refilling of the Black Sea did not exhibit catastrophic dynamics. During the last glacial maximum about 20,000 years ago, the Black Sea was an inland sea that increasingly began to become saline due to evaporation losses. Since about 17,000 years ago, an inflow from the Caspian Sea reached the basin via the Manych Depression and first rapidly, then gradually, raised the water level from −100 to −20 m. This resulted in an overflow into the Marmara Sea. A spillway into the Sea of Marmara and the Mediterranean Sea formed, whose water levels were still significantly lower than in the Black Sea Basin. Evidence of this key event is seen in the mollusc fauna, which can be distinguished between freshwater, brackish and saltwater species. Periodic changes in the species spectrum suggest changing dominant flow directions in the exchange between the Black and Mediterranean Seas from 9500 years before today's bidirectional system with light lower-salinity water at the surface from the Black Sea into the Mediterranean and higher-salinity water below with reversed flow direction. Water level changes in the Black Sea have not exceeded rates of 3 cm per century, which meant no noticeable changes for the settlers of the time. Catastrophic components that would allow a link to the Flood myth cannot be discerned in the history of the refilling of the Black Sea and are not confirmed by scientific investigations. The late glacial rise clearly exceeded the Holocene rise in intensity, but remained without verifiable archaeological consequences. A Holocene flood catastrophe is characterised as a myth in its own right without reference to reality or the biblical Flood.

6.1.4 Continuous Refill with Permanently Dominant Flow into the Mediterranean

This interpretive approach is mostly similar to the above, but sees evidence of a permanent flow southwards through the Bosporus for the last 10,500 years (Hiscott et al. 2007). The main argument are corresponding deposits in the Sea of Marmara and the adjacent Aegean. According to the authors' argumentation, this direction of flow does not exclude the migration of molluscs from the Mediterranean into the Black Sea, so that corresponding species spectra lose weight as an argument. The same applies to the strontium isotope spectrum, whose change is attributed to a temporal delay in signal transfer. With vividly processed sonar images, Flood et al. (2009) document that the geomorphological structures at the shelf margin of the Black Sea can be related to recent density currents at the bottom. The recent deep current from the Mediterranean into the Black Sea, which developed continuously and gradually since 7500 years before today, is quite capable of producing structures similar to those of high-energy outburst floods.

6.2 Largest Flood in Earth's History: Refill of the Mediterranean

In the 1970s, drilling by the research vessel Glomar Challenger yielded sensational results: The Mediterranean Basin (Fig. 6.3) had dried up around 6 million years ago (Hsü et al. 1973; Ryan 2009). Evidence was provided by thick and extensive salt deposits just below the seabed. The bedding structures and surface forms indicate that the salts were chemically precipitated in salt pans under hot and dry climatic conditions. How could it happen that all water of an ocean basin evaporated?

An important role is played by the fact that the water balance of the Mediterranean is negative, i.e. more water evaporates than flows in. In figures, this means for today's climate conditions that 4700 km^3 of water evaporate annually, but only 1230 km^3 are added from precipitation and 200 km^3 from the inflow of rivers such as the Ebro, Rhone, Po and Nile. In addition, the water balance of the Black Sea is already negative (cf. Sect. 6.1), so that the rivers flowing into it cannot contribute anything to the water balance of the Mediterranean. The annual net balance of −3270 km^3 would result in the Mediterranean Sea with a volume of 3,700,000 km^3 falling completely dry in about 1000 years. Of obvious importance is the inflow from the Atlantic through the Strait of Gibraltar, which alone provides the volume balance in the Mediterranean Sea. In order for this inflow to be stopped as well, the Strait of Gibraltar must become a land bridge between Europe and Africa, which was the case about 6 million years ago.

The closure of the strait between Europe and Africa is easily imaginable in the context of plate tectonic dynamics. It is well known that the African Plate is drifting northwards towards the European Plate, which has led to a narrowing of the space between the two and the rising of the young fold mountains. In the vicinity of the Strait of Gibraltar, these are specifically the Betic Cordillera in southern Spain and the Rift Mountains as an extension of the Atlas in Morocco. The ongoing discussion about tectonic sub-processes is not considered in detail here (cf. e.g. Duggen et al. 2003; Jolivet et al. 2006; Govers 2009), but it is important to visualise the different land-sea distribution compared to today at the time after the closure of the Strait of Gibraltar. Even though the recent and not the late-tertiary bathymetry is presented, Fig. 6.3 illustrates the separation of the deeper areas of the Mediterranean Basin into a western and an eastern part, between which the Sicilian Sill is located. During the Messinian time period of the late Tertiary, the Apennines had not yet been folded up and the Italian Boot therefore did not exist in its recent shape. Also, parts of the

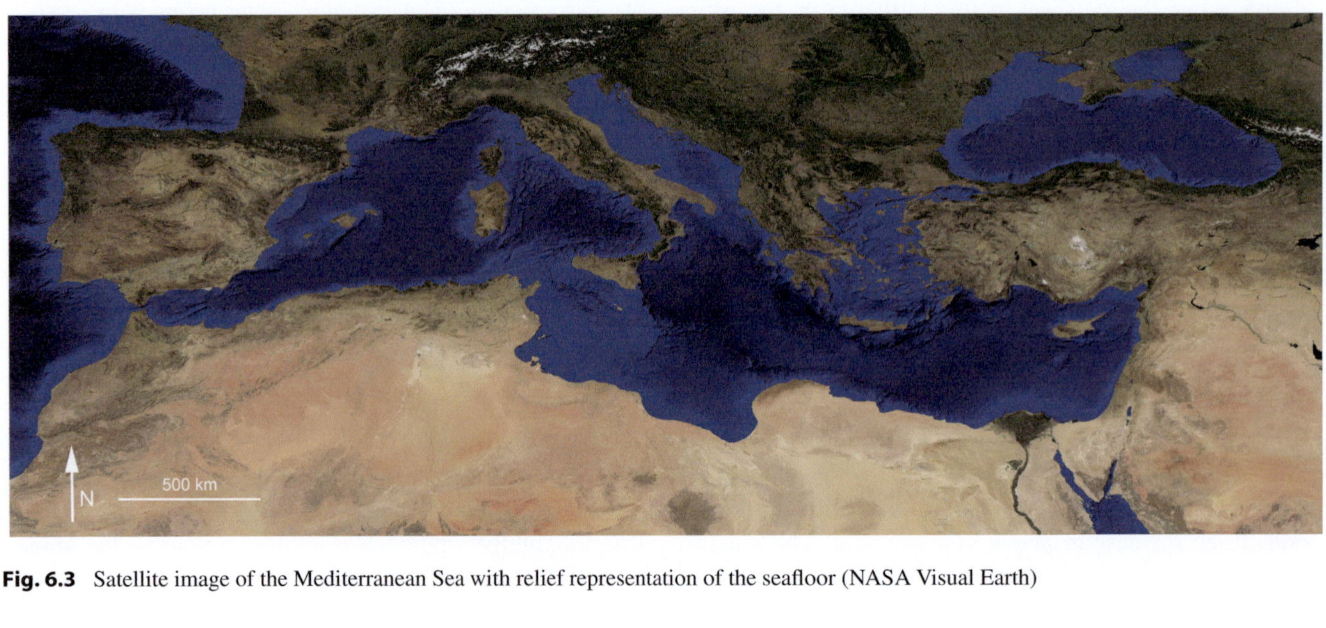

Fig. 6.3 Satellite image of the Mediterranean Sea with relief representation of the seafloor (NASA Visual Earth)

southern Mediterranean were still mainland and the Aegean had not yet opened up. As a result of the sinking sea level in the Mediterranean, the rivers flowing into it incised correspondingly deep, but their valleys filled up again when the water level rose again (Cita and Ryan 1978). This explains the overdeepened valleys and basins, for example along the southern margin of the Alps, which were cut to depths of 500 m below sea level (Felber and Bini 1997). Corresponding findings are also available, for example, for the River Ebro (Urgeles et al. 2010), southern France (Clauzon 1978) and the River Nile, where even incision depths of more than 2.5 km are observed (Barber 1981). The salt deposits with a volume of about 1 million km^3 have a thickness of 1500–3500 m (Meijer and Krijgsman 2005). These deposits are larger than the amount of dissolved salt contained in the Mediterranean water at maximum filling, so there must have been repeated desiccations, hence repeated so-called Messinian Salinity Crisis. This conclusion is confirmed by the observation of the bedding of the salt deposits. The sedimentary structures document that the salt was precipitated on a free surface and not from a supersaturated solution. The intermediate layers show a typical pattern of deep-sea deposits, so that the basin must also have refilled several times. The cause of these cycles is seen as periodic fluctuations in the water level of the Atlantic, which temporarily crossed a sill in the western basin margin (Meijer and Krijgsman 2005), but without the crossing water incising into the sill.

This incision occurred 5.33 million years ago when the Mediterranean Basin was replenished. The decisive turn was an incipient rift valley rupture that facilitated the overflow of water from the Atlantic (Govers 2009). The key to reconstruct the flow in the Strait of Gibraltar that was formed at that time is given by the topography of the seafloor (Fig. 6.4), although reconstructing the topography nearly 6 million years ago in detail is a challenge.

The topography of the seafloor east of the Strait of Gibraltar consists of a basin structure that already has depths of more than 1000 m, so that the application of modelling approaches of rivers in high mountains is appropriate (Garcia-Castellanos et al. 2009). After an initial overflow, the channel began to incise rapidly, causing the discharge to increase in a self-reinforcing manner. Incision rates exceeded 0.4 m per day and the discharge of up to 120,000,000 m^3/s reached flow velocities of typically 40 m/s. Due to the separation of the Mediterranean into several sub-basins with overflow sills, refilling was a multi-phase process that will have taken several years in total even if sea level rise rates of up to 10 m/day have to be expected (Periánez and Abril 2015; Garcia-Castellanos et al. 2020). However, modelling showed that 90% of the Mediterranean was refilled in less than two years. This discharge is orders of magnitude higher than any other flood event in Earth's history and is not even remotely approached by historical or recent events (cf. review in Chap. 1). For comparison: in recent times, almost 1,700,000 m^3/s flow constantly through the Strait of Gibraltar, which is notorious for its powerful current, with a typical flow velocity of 1–2 m/s.

6.3 The Great Deluge—Myths and Background

Flooded landscapes after heavy rainfall as far as the eye can see—the description is already present in the media and frequently characterized as torrential respectively diluvial. Even though it does not seem to be investigated in detail so

Fig. 6.4 Satellite image of the Strait of Gibraltar with relief representation of the seafloor (NASA Visual Earth)

far, probably the first written comparison of a recent flood event with be biblical deluge are handed down from Paul the Deacon, a Lombard historian, who wrote in the end of the eighth century about a flood event on 23rd of October 589 witnessed by the then pope (cf. Fontana et al. 2020 for further details). However, the biblical Flood has not only been used intensively on a daily basis and as a pictorial comparison, there have also been renowned geoscientists who wanted to make contributions to research on the geo-scientific background of the event: Alexander Tollmann (geology of the Alps), Cesare Emiliani (sea level changes), Herbert Liedtke (Nordic Glaciations in Central Europe) or William Ryan (desiccation of the Mediterranean) to mention some of them publishing their conclusions in more or less recent times. Details on these approaches are given and discussed below.

So was the Flood more than a myth? By geoscientists, searching for indicators in the field was started in the early nineteenth century, when in the age of the theory of Neptunism all rocks were assumed to be deposited by a primordial ocean and this reflected the biblical Flood (Montgomery 2013). Thus, according to this concept, it is quite consistent if boulders identified today as being of glacial origin are attributed to these diluvial (sic!) deposits by famous geologists. For corresponding findings in the US state of Maine, a water depth of the Flood of over 1500 m and an inflow direction from the northwest have been concluded (Jackson 1839, cf. Hineline (1988) for a historical analysis on the context). In his early years, Leopold von Buch gave a lecture on 31st of October 1811 and came to the conclusion that similar erratic boulders in northern Germany flew over the Baltic Sea with great force and high velocity. He quantified transport velocities of 170–6100 m/s for corresponding deposits in the Swiss Jura (von Buch 1815).

On the other hand, historicity of the Flood is frequently rejected, either as a postulate (e.g. Sonnabend (2013, p. 34) in his review on catastrophes in antique times) or because floods of the dimension described are expected as physically not possible (e.g. Lenzen (1964, p. 64) in his geoarchaeological analysis of flood deposits in Mesopotamia). Nowadays, there are different approaches applied in this context by geoscientists to take the traditions of the Flood myth literally and to regard them as historical facts, which is judged and discussed very controversially as examples illustrate below.

When entering the keyword "deluge" into internet search tools, the high number and enormous variety of contributions makes it obvious how popular the question of the origin of the Flood myth still is. Already in the 1950s, Keller (2009, p. 57) was able to draw on 80,000 publications on the subject in 72 languages. There are frequently headlines that the historic Flood can be fixed in time or that remains of Noah's Ark have been found. This is also evident from the supposedly sales-promoting effect of the term in book titles, as a glance at the bibliography at the end of the article illustrates.

6.3.1 Background of the Biblical Deluge

George Smith caused a great stir in 1872 when he succeeded in deciphering the Akkadian cuneiform script of the Epic of Gilgamesh from Mesopotamia (Dundes 1988, 29f): quite obviously, this epic, written around 1200 BC, and the Old Testament have a common Flood tradition (Table 6.1). That means that the Old Testament from about 700–400 BC is no longer an original text (Baumgart 2005). The Epic of Gilgamesh (Zgoll 2007) is an epic poem on 12 clay tablets and is about Gilgamesh, who was a legendary ruler and hero—part god, part man. His supreme wisdom comes from pre-Flood experiences gained through conversations with Utnapishtim, Noah's counterpart and first-person narrator in the epic. Since 2600 BC, he has been part of the Sumerian list of kings, which differentiates between a period before and after the Flood. It is unclear whether there is a historical person behind the epic.

Even though the chronology of freeing a raven respectively a dove differs in the compared texts, the parallels between the biblical narrative and the Epic of Gilgamesh are obvious: according to divine decree, a far-reaching destruction is planned by a flood resulting from abundant rainfall, but individuals (Noah or Utnapishtim) are saved by the reference to the construction of a flood-proof vehicle. The offerings made after salvation satisfy, and there is no repetition of an all-encompassing devastating flood.

The picture of the obvious adoption of older traditions in the Old Testament is complicated by the fact that components of the Old Babylonian Atra-Chasis epic or a Sumerian flood account from the 3rd millennium BC can also be found, but these are missing in the Epic of Gilgamesh (Baumgart 2005, 16ff). Apparently, the Flood motif was widespread in the Fertile Crescent cultural zone, which stretches from the ancient civilizations of Mesopotamia in an arc over Assyria and Phoenicia to the lower reaches of the Nile, and has thus found its way into various traditions over a longer period of time. This also includes the ancient Greek cultural area, where the Babylonian text was also known (Montgomery 2012, 154f) and was woven into the legendary tangle of legends written around 100 BC (Caduff 1986; Knauss 1987).

One remaining challenge is to explain the spread of the myth of exceptional large and destructive flood event in regions that were without contact with the Fertile Crescent or Greco-Roman antiquity. Corresponding motifs can be found in Celtic tales, among the Germanic peoples, from India, China, Thailand, the Philippines, Australia, southern

Table 6.1 Fading of the Flood in the Epic of Gilgamesh (Ferry 1991) and the Old Testament

Epic of Gilgamesh XI, 145–162	Genesis 8, 4–22
… and then the boat was caught and held from under by the peak rock of a mountain under the water. It was Mount Nisir the boat was grounded on	… and in the seventh month, on the seventeenth day of the month, the ark came to rest on the mountains of Ararat
… and then on the seventh day I freed a dove. The dove flew free and flew away from the boat, seeking a place for its little feet to alight, and finding none, flew back to the boat to perch	At the end of forty days Noah opened the window of the ark that he had made and sent forth a raven. It went to and fro until the waters were dried up from the earth
I freed a raven then and it flew free and flew away from the boat, and never returned. It had found a place to alight, and circled about the place, and alighted, and settled itself, and ate, and never after that returned to the boat	He waited another seven days, and again he sent forth the dove out of the ark. And the dove came back to him in the evening, and behold, in her mouth was a freshly plucked olive leaf. So Noah knew that the waters had subsided from the earth. Then he waited another seven days and sent forth the dove, and she did not return to him anymore
I went ashore and offered a sacrifice. I poured out a libation; I set out seven vessels of offerings on a stand, and then set seven more; I made a fire of wood of myrtle, wood of cane, and wood of cedar. I lit the fire. The odor touched the nostrils of the Igigi gods and gave their nostrils pleasure	Then Noah built an altar to the Lord and took some of every clean animal and some of every clean bird and offered burnt offerings on the altar. And when the Lord smelled the pleasing aroma, …

Note The chronology of freeing a raven respectively a dove differs in the text sources

Africa and the Indian cultures of America, whereby influence by Christian missionaries is taken into account (Baumgart 2005, p. 20; Caduff 1986, p. 11f; Dundes 1988).

6.3.2 Geoscientific Approaches of Interpretation of the Myths

A first attempt to decipher the natural background of the Flood myth is to take the traditional texts literally. This has been attempted by early geoscientists (cf. above) and has since been developed by creationists into a biblical-geological model in which the Flood is divided into several rising and falling phases totalling 370 days (Walker 1994). However, contradictory information calls for further interpretations: e.g. according to Genesis 7:17 the flood lasted 40 days, but according to Genesis 7:24 and 8:3 it lasted 150 days. Differing indications of the duration of the introductory rain—6 days in the Epic of Gilgamesh, 7 days according to the oldest Sumerian account, 9 days in the Greek sagas, and 40 days in the Old Testament—are not surprising. The figures, which express extraordinary long-lasting rain, are of regional and religious significance rather than hard facts (Knauss 1987).

It is also obvious to look for deposits of the flood in the source area of the oldest traditions: in Mesopotamia. Woolley's (1952) findings of a flood layer, which were found during excavations in the historic city of Ur, caused a corresponding sensation. The finding is contradicted with reference to a large number of flood deposits of different ages, composition and distribution, and other influencing factors such as tectonic adjustments, altered sediment load rates and underestimation of floods are brought into very controversial discussion (e.g. Lenzen 1964; Mallowan 1964; Raikes 1966). Baumgart (2005, 20) considers the connection between the archaeologically proven floods of different ages (e.g. Mallowan 1964; Brückner 2003) and the age of the ancient Near Eastern traditions to be speculative. He draws attention to the problem of the implausible global dissemination of the myth from this space.

This aspect is the focus of Tollmann and Tollmann (1993), who assume a breaking meteorite with several tsunamis resulting from impacts into the oceans as the cause of the flood. Following the motif of the extinction of the dinosaurs at the end of the Mesozoic Era 60 million years ago due to the impact in the Gulf of Mexico, their concept has the problem that no coherent evidence in the field could be found (Deutsch et al. 1994).

Galanopulos (1960), who refers to the Greek legends and assumes the eruption of the Santorini volcano around 1500 BC as the trigger, also considers tsunamis as the phenomenon behind the myth. The eastern Mediterranean is known as the site of repeated tsunamis and Bruins et al. (2008) were able to detect associated deposits and determine a wave height of around 10 m, which undoubtedly affected large parts of the Mediterranean coasts. Furthermore, the approach is integrated into a larger historical-cultural context—including a connection to the Atlantis myth—but this is sometimes too young for the ancient Near Eastern tradition and the question of the global spread of the myth is raised again.

On the other hand, the rise of the sea level after the last Ice Age (Liedtke 1958) (Fig. 6.5) is a global phenomenon, with all dry coastal shelfs being flooded over large areas, but also permanently, and still are today. This permanence does not contradict the Flood traditions, because the ships with the survivors were drifted and ran aground at higher altitudes (Table 6.1). The period is more problematic, because already around 6000 BP/4000 BC the current level had been already nearly reached and no rapid rates of

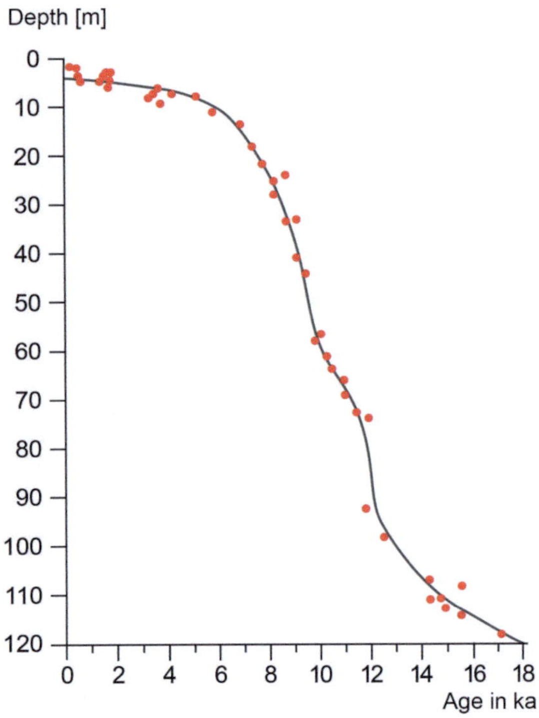

Fig. 6.5 Global sea level rise since the last ice age (modified according to various sources from Herget 2019)

increase could be recorded. Contrary to what Emiliani et al. (1975) interpreted, the speed of max. 1 cm/a was not really fast for human observers on the beach and therefore hardly catastrophic. On the other hand, it was certainly a traumatic experience for the people settling on the shelf areas on the coasts worldwide—as any visitor to extensive tidal flats e.g. on the North Sea coast can well understand at low tide.

Ryan and Pitman (1998) have developed a scenario according to which the transfer of water from the Mediterranean into the Black Sea as part of post-glacial global sea level rise led to a late and extremely rapid filling of the then inland sea. This approach is considered to be refuted, as there are also indications of opposing flow directions from Siberian ice-dammed lakes and thus into the Mediterranean Sea, while archaeological evidence seems to be lacking (cf. Sect. 6.1 for details).

6.3.3 Key Region of Mesopotamia and the Persian Gulf

Observations by Teller et al. (2000) focus on the Persian Gulf: large parts of the sand of the Arabian Peninsula consist of shell fragments that were blown out of the Persian Gulf during the last glacial period of a lowered sea level. Due to the flat topography, the sea level rise there was far-reaching despite the low rates, and around 5000 BP/3000 BC it temporarily reached a level about 1 m higher than today (Lambeck 1996; Lokier et al. 2015) (Fig. 6.6). Furthermore, Sarnthein (1972) and Lambert (1996) were able to demonstrate pronounced coastal structures on the seafloor, which indicates a phased flooding of the sea basin. Even without archaeological evidence, it must be assumed that the banks of the rivers running through the sea basin were first inhabited by man, because the climate there was even more arid at that time than it is today (Sarnthein 1972; Beni et al. 2024). While on the one hand, the sea advanced, on the other, the sediment-rich Rivers Euphrates and Tigris filled up a rapidly advancing delta, which explains why large parts of Mesopotamia today lie only slightly above or even below sea level. At the same time, the river courses themselves exhibit high dynamics with abrupt shifts and natural levee formation, which leads to large-scale flooding with corresponding deposits in case of a failure of the levees (Sherratt 2004; Montgomery 2012, p. 153).

The outlined combination of different influencing factors in the area of Mesopotamia and the Persian Gulf can explain many conflicts of other interpretative approaches (cf. Brückner and Engel 2019):

- The context and background of the Flood myth in the Epic of Gilgamesh, the Old Testament and Greek mythology find an origin (temporary risen sea level in the low-land area) and a coherent chronology
- Different flood deposits of the Rivers Euphrates and Tigris in Mesopotamia are placed into context
- Global post-glacial sea-level rise justifies the spread of the experience of extensive and habitat-destroying flooding beyond cultural connections and are not connected with the biblical Flood and geoarchaeological evidence in Mesopotamia.

Details and the weighting of the factors will certainly continue to be discussed controversially, but the advantages of a geographical-interdisciplinary approach are becoming obvious, as has already been shown by similar questions about the background of other traditions and legends (Vitaliano 1973; Keller 2009). In the meantime, Buchner and Buchner (2011) have already gone one step further and, on the basis of existing dating of coastal courses and flood deposits, have developed a chronology of settlement displacements in Mesopotamia as well as concluded the type and extent of individual floods, which can be viewed critically in view of dating uncertainties and the lack of validation of findings.

Finally, it should be pointed out that there are also completely different interpretations of the Flood myth without any context to hydrological floods: for example, Rank (in Dundes 1988, 151) interprets Freudian-psychoanalytically that the suppressed urge to urinate at night in one's sleep is the origin of the Flood myth, which easily explains its global distribution.

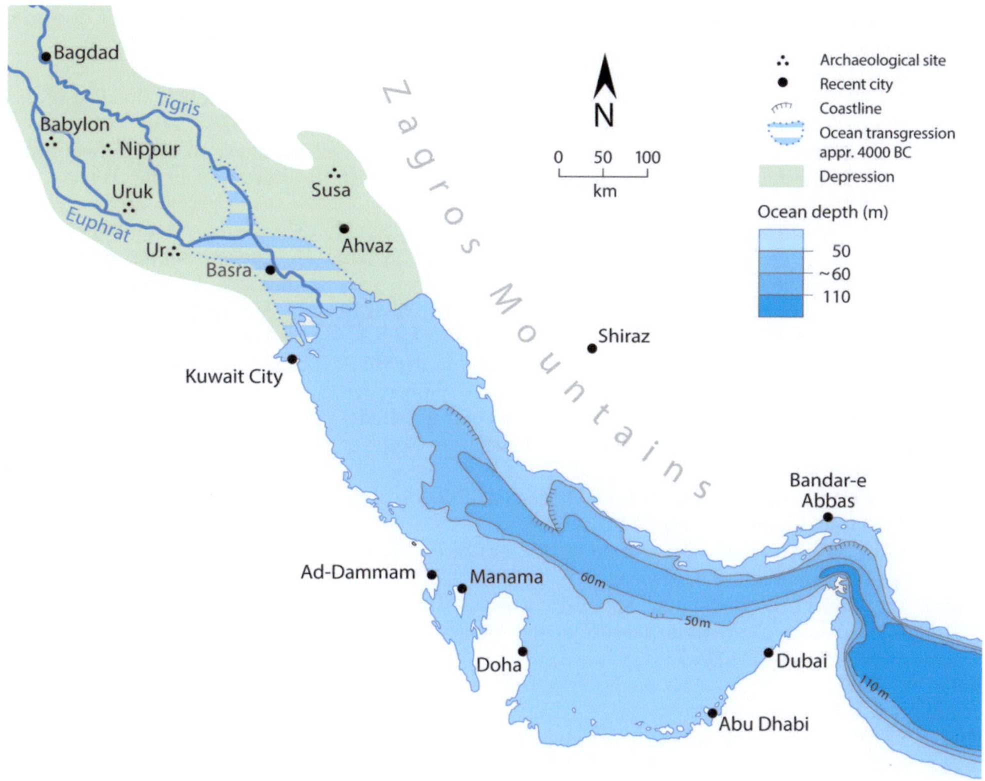

Fig. 6.6 Bathymetry of the Persian Gulf and sketched extent of the temporary flooding of Mesopotamia (modified after various sources from Herget 2019)

References

Barber PM (1981) Messinian subaerial erosion of the Proto-Nile delta. Mar Geol 44:253–272

Baumgart NC (2005) Sintflut/Sintfluterzählung. Wissenschaftliches Bibellexikon der Deutschen Bibelgesellschaft: http://www.bibel-wissenschaft.de/stichwort/11920. 9 May 2024

Beni AN, Leduc G, Djamal M, Sharifi A, Marriner N, Tachikawa K, Rostek F, Tjallingii R, Lahijani H, Arabshahi MM, Garcia M, Licari L, Tetard M, Bellinghery M-C, Bard E (2024) Postglacial flooding and Holocene climate shifts in the Persian Gulf. J Quat Sci. https://doi.org/10.1002/jqs.3614

Brückner H (2003) Uruk—a geographic and palaeo-ecologic perspective on a famous ancient city in Mesopotamia. Geoöko 24:229–248

Brückner H, Engel M (2019) Noah's flood—probing an ancient narrative using geoscience. In: Fontana A, Herget J (eds) Palaeohydrology—traces, tracks and trails of extreme events. Springer, Cham, pp 135–151

Bruins HJ, MacGillivray JA, Synolakis CE, Benjamini C, Keller J, Kisch HJ, Klügel A, van der Plicht J (2008) Geoarchaeological tsunami deposits at Palaikastro (Crete) and the Late Minoan IA eruption of Santorini. J Archaeol Sci 35:191–212

Buchner N, Buchner E (2011) Klima und Kulturen - die Geschichte von Paradies und Sintflut. BAG, Weinstadt

Buynevich IV, Yanko-Hombach V, Gilbert AS, Martin RE (eds) (2011) Geology and geoarchaeology of the Black Sea region—beyond the flood hypothesis. Geological Society of America Special Paper 473

Caduff GA (1986) Antike Sintflutsagen. Vandenhoeck & Ruprecht, Göttingen

Chepalyga AL (2007) Late glacial great flood in the Ponto-Caspian basin. In: Yanko-Hombach V, Gilbert AS, Panin N, Dolukhanov PM (eds) The Black Sea flood question—changes in coastline, climate and human settlement. Springer, Dordrecht, pp 119–148

Cita MB, Ryan WBF (1978) Messinian erosional surfaces in the Mediterranean. Mar Geol 27:193–366

Clauzon G (1978) The Messinian Var canyon (Provence, southern France)—paleogeographic implications. Mar Geol 27:231–246

Deutsch A, 12 co-authors (1994) The impact-flood connection—does it exist? Terra Nova 6:644–650

Duggen S, Hoernle K, van den Bogaard P, Rüpke L, Morgan JP (2003) Deep roots of the Messinian salinity crisis. Nature 422:602–605

Dundes A (ed) (1988) The flood myth. University of California Press, Berkeley

Emiliani C, Gartner S, Lidz B, Eldridge K, Elvey DK, Chang Huang T, Stipp JJ, Swanson MF (1975) Paleoclimatological analysis of Late Quaternary cores from the northeastern Gulf of Mexico. Science 189:1083–1088

Felber M, Bini A (1997) Seismic survey in alpine and prealpine valleys of Ticino (Switzerland)—evidences of a Late-Tertiary fluvial origin. Geologia Insubrica 2:47–67

Ferry D (1991) Gilgamesh—Tablets X and XI. Arion J Hum Class 1(3):93–116

Flood RD, Hiscott RN, Aksu AE (2009) Morphology and evolution of an anastomosed channel network where saline underflow enters the Black Sea. Sedimentology 56:807–839

Fontana A, Frassine M, Ronchi L (2020) Geomorphological and geoarchaeological evidence of the Mediaval Deluge in the Tagliamento River, NE Italy. In: Herget J, Fontana A (eds) Palaeohydrology—traces, tracks and trails of extreme events. Springer, Cham, pp 97–116

Galanopoulos GA (1960) On the origin of the Deluge of Deucalion and the myth of Atlantis. Greek Archaeol Soc 3:227–231

Garcia-Castellanos D, Estrada F, Jiménez-Munt I, Gorini C, Fernàndez M, Vergés J, De Vicente R (2009) Catastrophic flood of the Mediterranean after the Messinian salinity crisis. Nature 462:778–781

Garcia-Castellanos D, Micallef A, Estrada F, Camerlenghi A, Ercilla G, Perianez R, Abril JM (2020) The Zanclean megaflood oft he Mediterranean—searching for independent evidence. Earth Sci Rev 201:103061

Govers R (2009) Choking the Mediterranean to dehydration—the Messinian salinity crisis. Geology 37:167–170

Herget J (2019) Die Sintflut – Mythos und Realität. Geogr Rundsch 7(8–2019):52–57

Hineline M (1988) Charles Thomas Jackson and the first geological survey of Maine, pp 1836–1838. In: Maine Geological Survey (ed) Studies in maine geology, vol 1, pp 1–16

Hiscott RN, Asku AE, Mudie PJ, Kaminski MA, Abrajano T, Yasar D, Rochon A (2007) The Marmara Sea gateway since ~16 ky BP—non-catastrophic causes of paleoceanographic events in the Black Sea at 8.4 and 7.15 ky BP. In: Yanko-Hombach V, Gilbert AS, Panin N, Dolukhanov PM (eds) The black sea flood question—changes in coastline, climate and human settlement. Springer, Dordrecht, pp 89–117

Hsü KJ, Ryan WBF, Cita MB (1973) Late Miocene desiccation of the Mediterranean. Nature 242:240–244

Jackson CT (1839) Third annual report on the geology of the State of Maine. Smith and Robinson, Augusta

Jolivet L, Augier R, Robin C, Suc J-P, Rouchy JM (2006) Lithospheric-scale geodynamic context of the Messinian salinity crisis. Sed Geol 188(189):9–33

Keller W (2009) Und die Bibel hat doch Recht. Ullstein, Berlin (engl. title: The Bible as history)

Knauss J (1987) Deukalion, Lykoreia, die große Flut am Parnaß und der Vulkanausbruch von Thera. Antike Welt 3:23–41

Lambeck K (1996) Shoreline reconstructions for the Persian Gulf since the last glacial maximum. Earth Planet Sci Lett 142:43–57

Lenzen HJ (1964) Zur Flutschicht in Ur. Baghdader Mitteilungen 3:52–64

Liedtke H (1958) Sintflutmythos und Wissenschaft. Urania-Universum 4:308–313

Lokier SW, Bateman MD, Larkin NR, Rye P, Stewart JR (2015) Late Quaternary sea-level changes of the Persian Gulf. Quatern Res 84:69–81

Mallowan MEL (1964) Noah's Flood reconsidered. Iraq 26:62–82

Meijer PT, Krijgsman W (2005) A quantitative analysis of the desiccation and re-filling of the Mediterranean during the Messinian Salinity Crisis. Earth Planet Sci Lett 240:510–520

Montgomery DR (2012) The rocks don't lie—a geologist investigates Noah's flood. W.W. Norton, New York

Montgomery DR (2013) Faith in floods—field and theory in landscape evolution before geomorphology. Geomorphology 200:9–19

Myers PG, Wielki C, Goldstein SB, Rohling EJ (2003) Hydraulic calculations of postglacial connections between the Mediterranean and the Black Sea. Mar Geol 201:253–267

Periánez R, Abril JM (2015) Computational fluid dynamics simulations of the Zanclean catastrophic flood of the Mediterranean, 5.33 Ma. Palaeogeogr Palaeoclimatol Palaeoecol 424:49–60

Raikes RL (1966) The physical evidence of Noah's Flood. Iraq 28:52–63

Ryan WBF (2009) Decoding the Mediterranean salinity crisis. Sedimentology 56:95–136

Ryan WBF, Pitman WC III (1998) Noah's Flood—The new scientific discoveries about the event that changed history. Simon & Schuster, New York

Ryan WBF, Pitman WC III, Major CO, Shimkus K, Moskalenko V, Jones GA, Dimitrov P, Görür N, Sakinç M, Yüce H (1997) An abrupt drowning of the Black Sea shelf. Mar Geol 138:119–126

Ryan WBF, Major CO, Lericolais G, Goldstein SL (2003) Catastrophic flooding of the Black Sea. Ann Rev Earth Planet Sci 31:525–554

Sarnthein M (1972) Sediments and history of the postglacial transgression in the Persian Gulf and northwest Gulf of Oman. Mar Geol 12:245–266

Shackleton NJ (1987) Oxygen isotopes, ice volume and sea level. Quatern Sci Rev 6:183–190

Sherratt A (2004) Environmental change—the evolution of Mesopotamia. ArchAtlas Version 5.0: https://www.archatlas.org/EnvironmentalChange/EnvironmentalChange.php. 9 May 2024

Sidall M, Pratt LJ, Helfrich KR, Giosan L (2004) Testing the physical oceanographic implications of the suggested sudden Black Sea infill 8400 years ago. Paleoceanography 19:PA1024. https://doi.org/10.1029/2003PA000903

Sonnabend H (2013) Katastrophen in der Antike. WBG, Darmstadt

Teller JTKW, Glennie NL, Singhvi AK (2000) Calcareous dunes of the United Arab Emirates and Noah's Flood—the postglacial reflooding of the Persian (Arabien) Gulf. Quatern Int 68–71:297–308

Tollmann A, Tollmann E (1993) Und die Sintflut gab es doch - vom Mythos zur historischen Wahrheit. Droemer Knaur, München

Urgeles R, Camerlenghi A, Garcia-Castellanos D, De Mol B, Garcés M, Vergés J, Haslam I, Hardman M (2010) New constraints on the Messinian sealevel drawdown from 3D seismic data of the Ebro Margin, western Mediterranean. Basin Res 23:123–145

Vitaliano D (1973) Legends of the Earth—their geologic origins. Indiana University Press, Bloomington

Von Buch L (1815) Ueber die Ursachen der Verbreitung großer Alpengeschiebe. In: Abhandlungen der Königlichen Preußischen Akademie der Wissenschaften zu Berlin 1804–1811. Realschul-Buchhandlung, Berlin

Walker TB (1994) Charts of the biblical geological model: http://biblicalgeology.net. 9 May 2024

Woolley CL (1952) Ur of the Chaldees—a record of seven years of excavation. Penguin, Harmondsworth (reprint)

Yanchilina AG, Ryan WBF, McManus JF, Dimitrov P, Dimitrov D, Slavova K, Filipova-Marinova M (2017) Compilation of geophysical, geochronological, and geochemical evidence indicates a rapid Mediterranean-derived submergence of the Black Sea's shelf and subsequent substantial salinification in the early Holocene. Mar Geol 383:14–34

Yanko-Hombach V, Yilmaz Y (eds) (2007) IGCP521—Black Sea-Mediterranean corridor during the last 30 ka—sea level change and human adaptation. Quaternary International, pp 167–168

Yanko-Hombach V, Smyntyna OV (eds) (2009) Quaternary history of the Black Sea and adjacent regions. Quatern Int 197:1–128

Yanko-Hombach V, Gilbert AS (eds) (2018) IGCP 610 Second Special Volume. Quaternary International 465A

Yanko-Hombach V, Gilbert AS, Panin N, Dolukhanov PM (eds) (2007a) The Black Sea flood question—changes in coastline, climate and human settlement. Springer, Dordrecht

Yanko-Hombach V, Gilbert AS, Dolukhanov P (2007b) Controversy over the great flood hypotheses in the Black Sea in light of geological, paleontological, and archaeological evidence. Quatern Int 167–168:91–113

Zgoll A (2007) Gilgamesch. Wissenschaftliches Bibellexikon der Deutschen Bibelgesellschaft: http://www.bibelwissenschaft.de/stichwort/19604. 9 May 2024

Further Reading

A review on the different interpretations for the Black Sea refill with related assumptions and conclusions is given in

Yanko-Hombach V, Gilbert AS, Panin N, Dolukhanov PM (eds) (2007) The Black Sea flood question—changes in coastline, climate and human settlement. Springer, Dordrecht

Flood myths in different cultures and times, and geo-archaeological approaches are compiled in the reviews by

Dundes A (ed) (1988) The flood myth. University of California Press, Berkeley

Brückner H, Engel M (2019) Noah's flood—probing an ancient narrative using geoscience. In: Fontana A, Herget J (eds) Palaeohydrology—traces, tracks and trails of extreme events. Springer, Cham, pp 135–151

7 Conclusion and Perspectives

The preceding chapters have shown how diverse the history of floods on Earth is. On the one hand, the presentations touched on the classification of current flood dimensions in the historical context and, on the other hand, also listed the attempts to locate the oldest flood description ever using the example of the Flood myth. Motivation to study previous floods can be triggered by scientific curiosity like in the case of the Flood myths or the desiccation and refill of the Mediterranean. Megafloods occurred during Pleistocene times and similar geomorphological structures could be identified on other planets (Baker et al. 2015). The oldest flood traces date back to the geological time period of the Precambrian, around 2 billion years ago, when the Earth's surface was still free of larger vegetation and alternating flash floods poured over the bare surfaces (Erikson et al. 2006).

For mankind, floods provided both hazard and chance. While the hazard of destructive erosion, inundation, and sediment cover is obvious, floods also provide benefits as humic topsoil eroded upstream provide fertile environments by its deposition on the floodplains downstream. Early settled down high cultures like in Mesopotamia or Ancient Egypt were initiated by learning to manage the risk of floods and cultivate the fertile colluvial soils (Macklin and Lewin 2015).

The main area of application of experiences of historic floods is flood forecasting. Here, information on historic flood occurrence and magnitudes can make a significant contribution (e.g. Lam et al. 2017). As indicated in the chapter on flood frequencies in historical times, the established methods of flood estimation have problems due to the ongoing climate change. From the measured values of the river gauges, one can only update and extrapolate the actual state while trends have to be adjusted. Advances in statistics were made in the meantime, e.g. by consideration of stochastic uncertainty and imprecision of data (Salinas et al. 2016), missing statistical independence of events (Volpi et al. 2015) or short time of records and unsystematic data (Lam et al. 2017). Anyhow, flood frequency analysis is not suitable for forecasting floods during system changes in the hydrological regime, because the models might run "out of scale" (Furlani and Ninfo 2015). On the other hand, floods from historical times offer a great advantage, because they already trace a range of climate changes that actually existed. The Baltic Sea, which repeatedly froze over completely during the Little Ice Age on the one hand, and the possibility of farming in southern Greenland in medieval times on the other, illustrate the range of environmental changes due to different climatic conditions during the last thousand years. Analogously, flood reconstructions can document a similar range, even if they occurred under climatic conditions that do not correspond to the current ones. The perspectives considering the unknown climate conditions of the future and related prospective flood magnitudes and frequencies are obvious: just use the historical past for a plausible range of possibilities. Note, that large changes in the flood regime might be already triggered by modest climatic changes (Knox 1993).

The benefits of such approach can be illustrated by the flood event in July 2021 in western Germany and neighbouring regions (Roggenkamp and Herget 2024). Exceptional intensive rainfall of widespread > 150 mm in three days resulted in a catastrophic flood with 134 victims along the River Ahr. The largest discharge recorded before the event in the village of Altenahr with measurements since 1946 reached 236 m^3/s, while the water level during the July 2021 flood exceeded the top of the measuring rod by several metres (Fig. 7.1). As all gauging stations along the valley were destroyed by the flood, their undersized dimension is obvious and illustrates the underestimation of flood magnitudes based on statistical analysis of the surveyed data only. In historic times, the flood event in July 1804 reached a discharge of a similar magnitude of ~ 1200 m^3/s but was not considered for flood hazard assessment.

The benefits of consideration of historic flood events in flood hazard assessments was already identified before.

J. Herget, *Flood Reconstruction*, Springer Natural Hazards, https://doi.org/10.1007/978-3-031-66815-9_7

Fig. 7.1 Rebuilt but still undersized gauging station at Altenahr on the River Ahr, western Germany. The July 2021 Flood reached the balcony of the house in the background and destroyed all gauging stations along the valley

Consequently, the European Union developed a directive on the assessment and management of flood risks (EU 2007). In chapter III, article 6 is mentioned, that also for floods with a low probability or extreme event scenarios, the flow velocity or relevant water flow has to be shown in flood hazard maps. Obviously, the consideration of extreme events in historic times is included in this context. By a closer look, most of the recently published flood hazard maps do not provide such information. The reason might be seen, that in the cited article of the directive, the constraint "as/where appropriate" is also mentioned. By experience, for extreme contemporary flood events, surveying flow velocity and discharge tends to be less accurate due to complex current conditions. Consequently, and as related to uncertainty, quantification of extreme historic flood parameters so far appeared less appropriate and was left out frequently beyond academic studies. The methods for flood reconstruction compiled in a review in the first part of this book might support filling this gap and the results could improve flood hazard estimations. Following the title of an article by Baker (2003) on the importance of knowledge of historical floods, one might therefore conclude with the prediction: "A bright future for old flows".

References

Baker VR (2003) A bright future for old flows—origins, status and future of paleoflood hydrology. In: Thorndycraft VR, Benito G, Barriendos M, Llasat MC (eds) Palaeofloods, historical data & climatic variability—applications in flood risk assessment. Madrid (Centro de Ciencias Medioambientales), pp 13–18

Baker VR, Hamilton CW, Burr DM, Gulick VC, Komatsu G, Luo W, Rice JW Jr, Rodriguez JAP (2015) Fluvial geomorphology on Earth-like planetary surfaces—a review. Geomorphology 245:149–182

Erikson PG, Bumby AJ, Brümer JJ, van der Neut M (2006) Precambrian fluvial deposits—enigmatic palaeohydrological data from the ca. 2–1.9 Ga Waterberg Group, South Africa. Sediment Geol 190:25–46

EU—European Union (ed) (2007) Directive 2007/60/EC of the European Parliament and of the Council of 23 October 2007 on the assessment and management of flood risks. Official Journal of the European Union L 288, pp 27–34

Furlani S, Ninfo A (2015) Is the present the key to the future? Earth Sci Rev 142:38–46

Knox JC (1993) Large increase in flood magnitude in response to modest changes in climate. Nature 361:430–432

Lam D, Thompson C, Croke J, Sharma A, Macklin M (2017) Reducing uncertainty with flood frequency analysis—the contribution of paleoflood and historical flood information. Water Resour Res 53. https://doi.org/10.1002/2016WR019959

Macklin M, Lewin J (2015) Rivers of civilisation. Quatern Sci Rev 114:228–244

Roggenkamp T, Herget J (2024) Flood reconstruction—the unexpected rather frequent event at River Ahr in July 2021. Global and Planetary Change. https://doi.org/10.1016/j.gloplacha.2024.104541

Salinas JL, Kiss A, Viglione A, Viertl R, Blöschl G (2016) A fuzzy Bayesian approach to flood frequency estimation with imprecise historical information. Water Resour Res 52. https://doi.org/10.1002/2016WR019177

Volpi E, Fiori A, Grimaldi S, Lombardo F, Koutsoyiannis D (2015) One hundred years of return period—strengths and limitations. Water Resour Res 51:8570–8585

MIX
Papier aus verantwortungsvollen Quellen
Paper from responsible sources
FSC® C105338

If you have any concerns about our products,
you can contact us on
ProductSafety@springernature.com

In case Publisher is established outside the EU,
the EU authorized representative is:
Springer Nature Customer Service Center GmbH
Europaplatz 3, 69115 Heidelberg, Germany

Printed by Libri Plureos GmbH
in Hamburg, Germany